Artificial Intelligence and Alloy Design

David J. Fisher

Published by **Materials Research Forum LLC**
Millersville, PA 17551, USA

Published as part of the book series
Materials Research Foundations
Volume 166 (2024)
ISSN 2471-8890 (Print)
ISSN 2471-8904 (Online)

Print ISBN 978-1-64490-314-8
ePDF ISBN 978-1-64490-315-5

Distributed worldwide by

Materials Research Forum LLC
105 Springdale Lane
Millersville, PA 17551
USA
http://www.mrforum.com

Printed in the United States of America
10 9 8 7 6 5 4 3 2 1

Table of Contents

Introduction

Given the number of elements in the periodic table, the development of alloys has always involved searching a space of high dimension. Even with guides such as the Hume-Rothery rules, the exploration of the vast combinatorial space is rather slow and expensive. The development of new materials has already experienced three historical stages: empirical, theoretical and computational. The huge volume generated by experiment and simulation has now facilitated a shift in materials science to a data-driven fourth stage: the development of high-throughput automatic computation and data-mining of material databases. Artificial intelligence is crucial in accelerating the design of novel materials. It is playing an increasing role in the field of alloy design, as it furnishes various techniques which can accelerate the process of developing new alloys possessing the desired properties. Artificial intelligence can be used to predict the properties of alloys, based simply upon their composition. Machine-learning methods such as regression and so-called deep-learning can analyze huge datasets of material properties and alloy compositions so as to establish significant patterns and thus predict how particular alloy compositions will behave. This can then significantly accelerate the identification of promising alloy candidates.

The AI-driven simulations can rapidly assess the properties of numerous alloy compositions without the need for extensive and expensive experimental testing. This narrows the search-space and identifies those compositions which are most likely to satisfy specific criteria. The AI tools, which include genetic algorithms and reinforcement learning, can optimize alloy compositions so as to improve, for example, the strength, corrosion resistance or thermal conductivity. The algorithms can iteratively generate and test alloy compositions; learning from each iteration and refining the search for the ideal composition. It is possible to reveal complex relationships between material properties, such as how changes in composition affect hardness in the presence of other interfering influences.

Novel alloy combinations may be proposed which humans might rule out on the basis of habit. High-entropy alloy combinations, for example, flouted the familiar Hume-Rothery rules (figure 1). These were not the fruit of artificial intelligence, but one can anticipate that similar breakthroughs may well result from the entirely unbiased search strategies of AI. By analyzing vast datasets and recognizing patterns, AI can propose unconventional compositions that have the potential to exhibit uniquely superior characteristics.

Materials Research Forum LLC
https://doi.org/10.21741/9781644903148

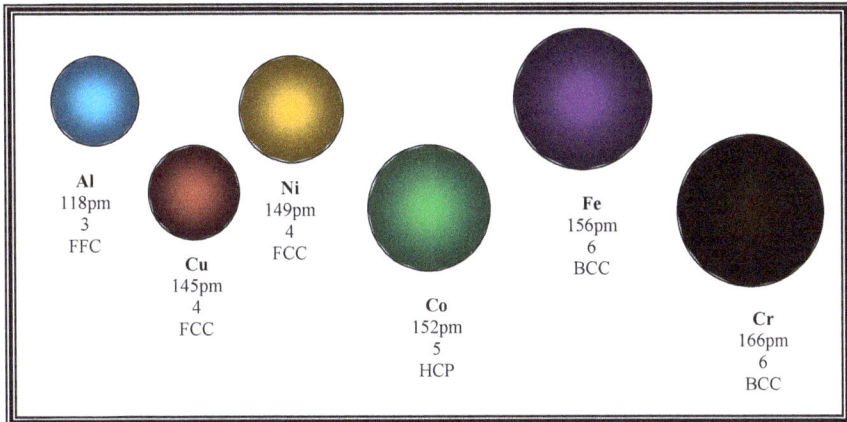

Figure 1. One can now envisage an alloy being made from near-equal fractions of the above elements, shown in order of size (but not to scale). It is obvious that such an alloy breaks Hume-Rothery's 15% rule for solid-solution formation. The valences and room-temperature structures also disagree, thus breaking Hume-Rothery's other rules

These same methods can further aid in the optimization of the manufacturing processes which are used to form components, and AI can monitor and control factors such as temperature and pressure during production as well as detecting defects or deviations from specifications. Following the proposal of candidate compositions, AI can further guide the efficient testing of alloy samples so as to minimize the number of experiments required. Looking forward, the future advent of quantum computing can serve only to revolutionize further the fields of materials science and alloy design.

The use of AI-type methods in alloy design has a surprisingly long history. As some workers bemoaned, a half-century ago, there lacked, "A systematic approach to the alloy designing problem, which is one of the typical ill-structured problems considered to be solvable only by inspirations of a genius metallurgist"[1]. One could argue that this was certainly true in the case of the much later discovery of high-entropy alloys[2]. At the earlier date it was reasoned that the procedure for designing an alloy possessing the required properties consists of two stages: the search for a starting-point (a so-called root), and the construction of a tree in order to improve the properties of the material which corresponded to the root by employing various tactics. A computer-aided design system with a metallurgical data-base for metallurgy might be used to establish a method for performing the procedure. New classes of alloy indeed tend to arise from chance discoveries or untargeted experiments, with improvements then being made via directed experimentation. Improvements which are guided by theoretical predictions follow later when empirical trends are recognised or when the relevant physical processes are better understood. At the end of the 20th century, artificial intelligence techniques were being explored as aids to the spotting of empirical trends within large databases. An improved understanding of theory, combined with dramatic increases in computational power led to models which could predict the basic physical properties of new compounds and alloys without experimentation. The electronic properties of exotic inorganic compounds which could not possibly be made, could nevertheless be predicted with confidence. Pattern recognition and neural network methods were used to investigate the formability of metastable alloy phases. It was found that chemical bond parameters such as valence electron number, electronegativity and metallic radii of the component elements are the predominant factors affecting metastable alloy phase formation. Semi-empirical rules which were found in that way could[3] be useful for the construction of expert systems for materials design. ALADIN, for example, was a knowledge-based system[4] which aided metallurgists in designing new aluminium alloys. Its development involved honing the representation of complex metallurgical knowledge. The system was a hybrid of several artificial intelligence techniques, with declarative representations being used to treat metallurgical concepts. The design was encoded within a rule base, with user interaction. The system was based upon a hypothesize-and-test model, and a constraint-based search was used to find alternative designs; with strategies included that resolved conflicts among competing sub-problems[5]. As well as the choice of alloy compositions themselves, alloy design must also consider the alloy's behaviour during processing and joining operations. The joining of aluminium-alloy 508 by DC spot-welding was studied[6] by using a discrete artificial neural network (ANN) model for quality prediction. An artificial neural network is a brain-like structure which learns by experience. It consists of an input layer, a hidden layer and an output layer. The input layer acts as a channel which passes information to the hidden layer by multiplying it with weights. The hidden layer applies a pre-defined mathematical model known as an activation function. The output of the hidden layer is passed to the output layer after multiplying it by further weights.

This showed that a reflection from input vector space to output vector space could be achieved via discrete processing of the input and output parameters of the model, such as welding-current, the voltage between electrodes and the shear strength. The predictive model offered good reliability and fault-tolerance. It is interesting to see how AI has been applied to aluminium alloys ever since.

Aluminium Alloys

A combined experimental and computational approach was used[7] for the accelerated design of Al-Ni-Co-type permanent magnet alloys. The concentrations of 8 alloying elements were initially generated by using a quasi-random number generator. Experimental data were used to develop meta-models that were capable of relating the chemical composition to desired macroscopic properties. The data were further used to identify correlations within the experimental data-set by using an unsupervised neural network and self-organizing maps. These maps were used to screen the compositions of a new generation of alloys to be manufactured and tested. Some of these Pareto-optimized predictions out-performed the initial batch of alloys. The approach significantly reduced the time and number of alloys required for alloy development.

Scandium is the best alloying addition for improving the mechanical properties of Al-Si-Mg casting alloys. Most work is aimed at choosing optimum scandium additions for various Al-Si-Mg casting alloys having well-defined compositions, but no attempt was made to optimize the contents of silicon, magnesium and scandium. This was due to the problem of exploring the high-dimensional compositional space by using a feasible number of experiments. A new alloy-design strategy was successfully used[8] to accelerate the discovery of hypoeutectic Al-Si-Mg-Sc casting alloys within such a high dimensional space. High-throughput calculations of phase diagrams (CALPHAD) solidification simulations were first made of hypoeutectic Al-Si-Mg-Sc alloys over a wide composition range in order to establish a quantitative relationship between composition, processing and microstructure. This relationship was derived by using active learning techniques, supported by key experiments which were proposed by CALPHAD and Bayesian optimization. Using a benchmark A356-xSc alloy, this strategy was used to design high-performance hypoeutectic Al-xSi-yMg alloys with optimum scandium contents, as later confirmed experimentally. This strategy was further successfully extended to explore the optimum contents of silicon, magnesium and scandium within the high-dimension hypoeutectic Al-xSi-yMg-zSc composition-space.

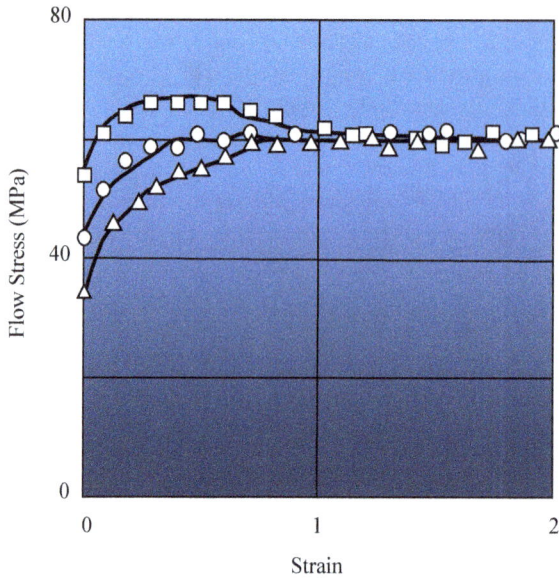

Figure 2. Comparison of experimental (points) and hybrid-model (lines) prediction results for the flow stress of a commercial-purity Al–1%Mg alloy following plane-strain compression at 385C under transient deformation conditions, after annealing at 400C. Squares: decreasing, circles: constant, triangles: increasing PSC test.

The composition of Al-Si cast alloys was studied by using microstructural image-recognition and machine-learning methods. Binary alloys with 1 to 10wt%Si were cast as reference images for machine-learning[9]. Repeated training related the microstructural images to the chemical composition, using up to 10000 steps. Peaks of similarity existed in the data-set for compositions which corresponded to known target compositions. The heights of the peaks became higher, and the distribution of similarity became sharper, with increasing number of training steps. The weighted average of the chemical composition approached the target composition with increasing training. The accuracy of the analysis increased with training steps of up to 10000 and then saturated. Compositions outside of the data-set could not be analyzed correctly. Analysis of the compositions between data-sets gave incorrect but reasonable results.

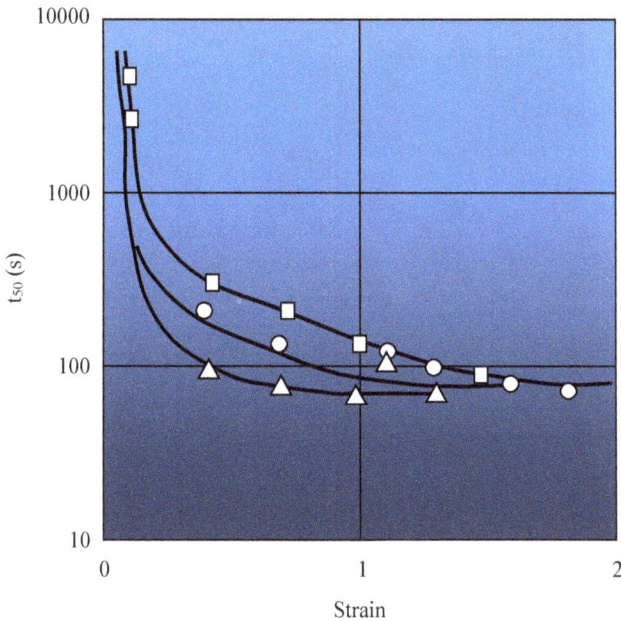

Figure 3. Comparison of experimental (points) and hybrid-model prediction (lines) results for the recrystallization kinetics of a commercial-purity Al–1%Mg alloy following plane-strain compression at 385C under transient deformation conditions, after annealing at 400C. Squares: decreasing, circles: constant, triangles: increasing PSC test.

An investigation was made[10] of the effect of a small (0.01at%) copper addition upon precipitation in an Al-0.80Mg-0.85Si alloy during ageing. The precipitate crystal structures were assessed by using scanning transmission electron microscopy and a scanning precession electron diffraction approach which incorporated machine-learning. This combination of techniques permitted evaluation of the atomic arrangement within individual precipitates, and of the fraction of precipitate phase for a given ageing condition. The total number of solute atoms locked within precipitates could be approximated. Even when the copper content was close to impurity-level, the precipitation was markedly affected by over-aging. The main change was due to a gradually increasing fraction of the Cu-containing phase, which eventually dominated the precipitate structures.

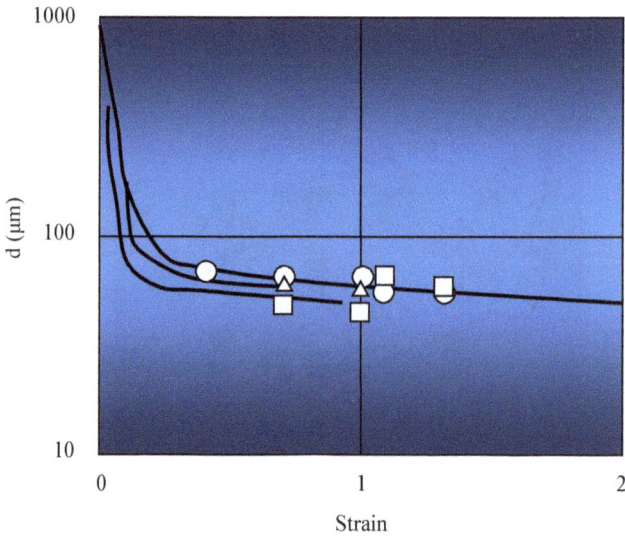

Figure 4. Comparison of experimental (points) and hybrid-model prediction (lines) results for the recrystallized grain-size of a commercial-purity A1–1%Mg alloy following plane-strain compression at 385C under transient deformation conditions, after annealing at 400C. Squares: decreasing, circles: constant, triangles: increasing PSC test.

A semi-physical hybrid model was developed[11] for the prediction of material behaviours such as the stress-strain relationship and recrystallization changes. The input variables were the temperature, strain-rate and strain. The model predicted the response to plain-strain compression (PSC) tests by using a hybrid combination of neuro-fuzzy intelligent models and constitutive equations. An integration process linked hybrid and finite-element models, with the hybrid model being embedded in the finite-element model and used to calculate the flow-stress distribution in PSC tests. The principal test-case was that of an Al-1%Mg alloy. The model predicted internal state variables as well as final material properties (figures 2 to 7). Internal state variables are particularly difficult and time-consuming to measure. The technique was applied to both 2-dimensional and 3-dimensional finite-element models for plane-strain compression tests. Empirical models are based upon mathematical expressions which fit the relationship between the input deformation conditions and the output variables. They are designed for fixed temperatures, strain-rates and compositions and do not model the internal states. The present hybrid modelling approach permitted the calculation of internal states: dislocation-density, sub-grain size and misorientation. The input deformation conditions could be extended to broader ranges of temperature and strain-rate, and used to treat various alloys. The technique could be extended to steels.

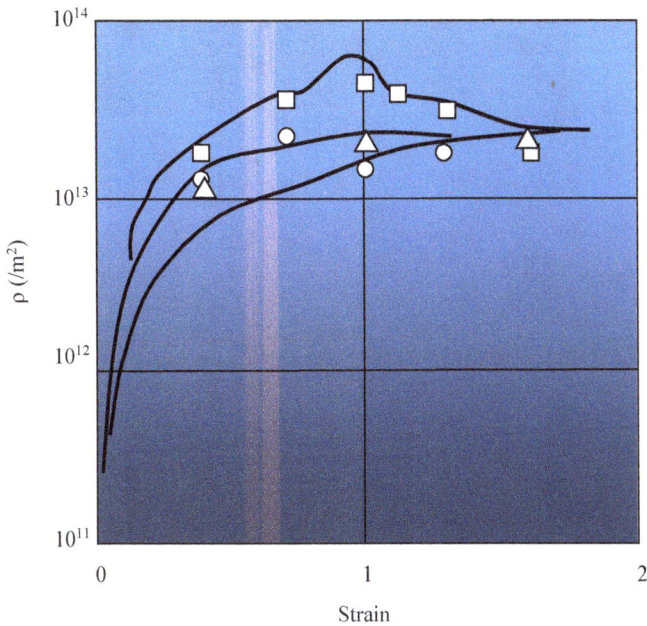

Figure 5. Comparison of experimental (points) and hybrid-model prediction (lines) results for the internal dislocation density of a commercial-purity Al–1%Mg alloy following plane-strain compression at 385C under transient deformation conditions, after annealing at 400C. Squares: decreasing, circles: constant, triangles: increasing PSC test.

A search was made[12] for new aluminium alloys, having improved mechanical properties, with the aid of artificial intelligence methods. A high prediction rate of AA7XXX-series aluminium alloys was achieved by using a Bayesian hyperparameter optimization algorithm. New alloys were designed which possessed an excellent combination of strength and ductility, such as a yield strength of 712MPa and an elongation of 19%. An homogeneous distribution of nanoscale precipitates hindered dislocation movement during deformation. The addition of magnesium and copper was the critical factor which determined the ratio of strength to elongation. In spite of the mechanistic interactions in the multiple steps of alloy fabrication, artificial intelligence successfully identified a new high-strength alloy by learning from experimental data.

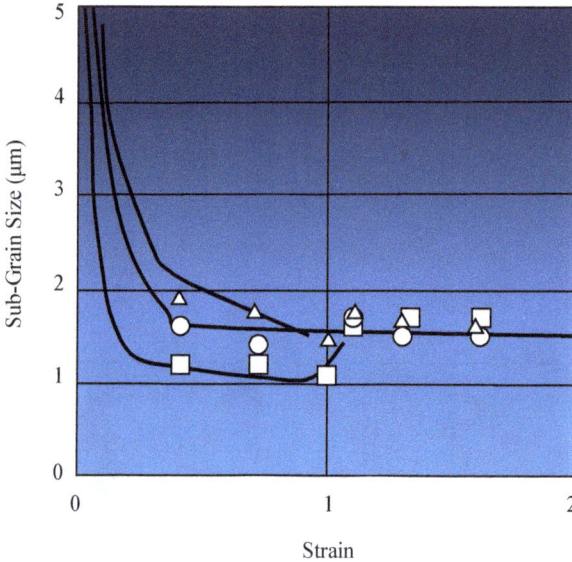

Figure 6. Comparison of experimental (points) and hybrid-model prediction (lines) results for the spacing between sub-boundaries of a commercial-purity A1–1%Mg alloy following plane-strain compression at 385C under transient deformation conditions, after annealing at 400C. Squares: decreasing, circles: constant, triangles: increasing PSC test.

There was a direct correlation between microstructure changes, caused by fabrication conditions, and the chemical composition. The resultant mechanical properties of the AA7XXX alloys could be used as input variables for deep-learning. A deep neural network was used to predict the mechanical properties of the series on the basis of 7 compositions and 5 fabrication processes. The optimum hyperparameter was determined by using a Bayesian optimization algorithm. By applying the K-fold cross-validation method, a direct neural network model of high predictive accuracy, with mean absolute percentage errors of 5.5830 and 9.9270, and coefficient of determination values of 0.9808 and 0.9260, in the training and test sets, respectively, using 227 small datasets. The local interpretable model-agnostic explanation (LIME) algorithm explained the positive or negative contribution of each variable to the mechanical properties, consistent with metallurgical theory. In the case of the positive contribution of zirconium, the element formed Al_3Zr. This is known to increase the yield strength and tensile strength as well as the elongation.

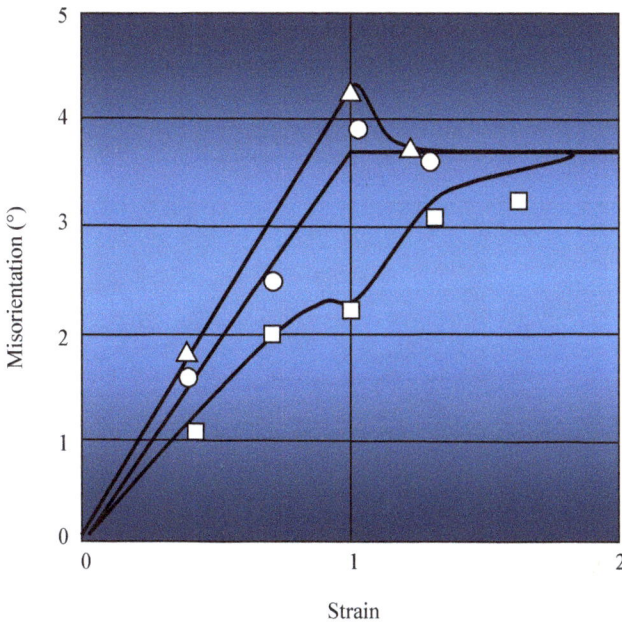

Figure 7. Comparison of experimental (points) and hybrid-model prediction (lines) results for the misorientation across the sub-boundaries of a commercial-purity Al–1%Mg alloy following plane-strain compression at 385C under transient deformation conditions, after annealing at 400C. Squares: decreasing, circles: constant, triangles: increasing PSC test.

Plastic forming and aging are known to have a large influence upon the improvement of mechanical properties. Although LIME could explain the independent contribution of each variable, there was a limit in that the correlated effect of specific elements or fabrication processes could not be clearly explained overall. It was possible to design an alloy having desired mechanical properties by using a recommendation algorithm to identify the optimum fabrication conditions. Microstructural analysis revealed that the recommended aging conditions had a significant effect upon the formation of nanoprecipitates, leading to strength enhancement. Experimental and simulation results were compared[13] for aluminium alloys containing from 0.5 to 13% of magnesium, plus smaller amounts of other elements. The AA5XXX alloys were made into sand-cast specimens. Those containing 12wt%Mg were subjected to tensile testing, and the tests were also computer-simulated by using artificial-intelligence techniques. The results of the computer simulations, and those of the tests, were found to be equivalent.

Table 1. Composition ranges within the clusters identified by machine learning

Group	0	1	2	3	4
Si	0.37–0.45	0.5–0.6	1–1.35	1–1.25	0.8–0.9
Fe	0.1–0.18	0.35–0.37	0.25	0.25–0.35	0.25
Cu	0.05–0.1	0.2–0.28	0.28–0.95	0.1–0.2	0.85
Mn	0.03–0.13	0.15	0.7–0.85	0.6–0.7	0.5–0.6
Mg	0.45–0.65	0.9–1	0.85–1.1	0.8–0.9	1–1.4
Yield strength	170–220	220–280	300–385	220–270	350–430
Toughness modulus	10.5–22	25–35	35.15–36.55	22–27	35.03–49
Mg:Si	1.3–1.68	1.667	< 1	< 1	1–1.667
Cu content	low	medium	high	low	high
Si excess	slight	stoichiometry	high	high	yes
Mg+Si+Cu	1.2 (low)	1.9 (medium)	2.9 (high)	2.1 (medium)	2.65 (high)

Incremental sheet-forming is a technique in which thin metal sheets are shaped with the aid of a simple tool. The process requires a simple fixture to grip the sheets. Artificial neural networks were used[14] to predict the deformation force for AA3003. The input variables included the feed-rate of the tool, the rpm of the spindle, the wall-angle, the

Materials Research Forum LLC
https://doi.org/10.21741/9781644903148

metal thickness and the density of lubricant. The recorded deformation forces ranged from 3.021 to 89.904N. The minimum deformation force corresponded to the deformation of 0.2mm-thick sheets at a step-downsize of 0.1mm. The overall coefficient-of-regression was 92.569% and the mean error was 5.878.

Conventional tests for determining the tensile strength of the alloys used in the aerospace industry were investigated[15] by digitization using the Levenberg-Marquardt algorithm, with the aim of investigating the performance of artificial neural networks as an alternative method for determining the tensile strength of metals. An analytical model was created by using a spreadsheet that was then digitized using artificial neural networks, leading to predictions of high accuracy.

As well as aiding the search for new alloys, artificial intelligence has also been used[16] to try to reduce the number of grades within the AA6XXX group of aluminium alloys. This involved collating features involving chemical composition and tensile properties in the T6 tempering state. It demonstrated the efficiency of grouping the alloys into clusters by using a combination of principal component analysis and the K-means algorithm, in which every data-point is repeatedly assigned to the cluster with the closest centroid and then, based upon the mean of the data points assigned to each cluster, the centroids or cluster centres are updated. Machine-learning algorithms identified a meaningful selection of 5 clusters (tables 1 and 2) which incorporated 50 commercial AA6XXX alloy grades. These served as a good basis for the optimization and reduction in the number of alloy grades, without impairing the tensile properties.

Table 2. Composition values affecting the stress and the toughness modulus

Cluster	Yield Stress	Composition Variation in Cluster*	Grades
4	highest	**Cu, Mg**,Si,Fe,Mn	AA6110A, AA6013, AA6092
2	2nd highest	**Si, Cu, Mn**, *Fe, Mg*	AA6024, AA6070, AA6066
1	medium	*Mn, Si*, **Fe**, Cu, Mg	AA6151, AA6053, AA6061
3	2nd lowest	**Si, Fe, Mg, Mn**, *Cu*	AA6005A, AA6009, AA6012
0	lowest	*Fe, Si, Mg, Mn, Cu*	AA6106, AA6101A, AA6060

*bold: high concentration, italic: low concentration, normal: medium concentration

The joining of alloys was also examined[17] by using fuzzy inference and fuzzy neural network (FNN) approaches, with parameters such as the electrical conductivity and yield strength being used as inputs. A FNN improved the intelligence of the design process and its learning ability. A data-base of numerical simulations of the resistance spot-welding aluminium alloys complemented the FNN training; enriching the training of the FNN and intensifying the generalizing abilities of the system.

A regression model, artificial neural network and adaptive neuro-fuzzy inference system were developed[18] for predicting the joint strength of stir melt welding of aluminium and copper sheets. The experiments were planned using a full factorial design process with 3 critical process parameters: the vibration amplitude, the weld pressure and the weld time. An analysis of variance study showed that the weld pressure has the greatest impact upon the tensile strength and the T-peel force, followed by the weld-time and the vibrational amplitude. The adaptive neuro-fuzzy inference system model offered an error of less than 1%.

The tensile strength, hardening and density of metal-matrix composites reinforced with micron-sized α-Al_2O_3 particles, and produced by stir-casting process, were predicted[19] by means of a back-propagation neural network that employed a gradient-descent learning algorithm. An artificial neural network solved non-linear problems by learning from the samples. Experimental samples were initially prepared in order to train the ANN and estimate the tensile strength, hardening and density of composites as a function of particle size. Following ANN training, using experimental samples, it furnished approximately correct outputs for experimental inputs that had not been used for training purposes. In experiments, 10vol% of particles of varying size was added to Al-Si-10Mg alloy by stir-casting. Mechanical tests showed that the tensile strength and hardness of the composites decreased with increasing particle size. The neural network was then trained by using the training set. Various particle sizes were used as the input, and the tensile strength, hardening and density were used as outputs in the neural network training module. The desired properties could eventually be estimated for various particle sizes by using the neural network rather than time-consuming experiments.

An artificial neural network model was sought[20] for predicting experimental flow stress in mechanically-alloyed AA6063-0.75Al_2O_3-0.75Y_2O_3 hybrid nanocomposites. The model was implemented using a feed-forward back-propagation network and log-sig functions, and a set of 80 training data and 20 testing data was chosen. The network comprised 3 input parameters (temperature, strain, strain-rate), a single hidden layer with 22 neurons and just a single output parameter: the flow stress. The latter was also predicted by using an Arrhenius constitutive model. Based upon a comparison of the predictions of the ANN model and of the Arrhenius constitutive model, it was concluded that the ANN offered a better accuracy and could be used to estimate the flow stress during the hot-deformation of the hybrid nanocomposite, with minimum error rates.

Table 3. R^2 values for various regression models and split-ratios

Regression Model	Split-Ratio	R^2(%)
AdaBoost	70:30	77.789
AdaBoost	80:20	82.061
AdaBoost	90:10	4.599
CatBoost	70:30	85.437
CatBoost	80:20	89.504
CatBoost	90:10	88.763
Decision Tree	70:30	78.630
Decision Tree	80:20	92.029
Decision Tree	90:10	76.389
Extra Tree	70:30	84.156
Extra Tree	80:20	89.349
Extra Tree	90:10	87.987
Gradient Boost	70:30	85.206
Gradient Boost	80:20	85.553
Gradient Boost	90:10	82.595
Random Forest	70:30	78.227
Random Forest	80:20	85.133
Random Forest	90:10	82.772
XGBoost	70:30	84.351
XGBoost	80:20	85.400
XGBoost	90:10	82.826

The AA2024 alloy was reinforced with alumina nanoparticles by means of multiple passes of friction stir processing. The effects of the multiple passes upon grain refinement and various mechanical properties were collated and used to train a hybrid artificial intelligence model[21]. The latter consisted of a multilayer perceptron which was driven by a grey-wolf optimizer so as to predict the mechanical and microstructural properties of friction stir processed aluminium alloy reinforced with alumina nanoparticles. The inputs of the model were the rotational speed, the linear processing speed and the number of passes, while the outputs were the grain-size, the aspect ratio, the microhardness and the ultimate tensile strength. The predictive accuracy of the hybrid model was compared with that of a standalone multilayer perceptron model, showing that the former offered a higher accuracy.

Table 4. Errors for various regression models (80:20 split) for the UTS of hybrid composites

Regression Model	Mean Absolute Error	Maximum Error	Root Mean Squared Error
Decision Tree	15.0179	43.10	19.577
Random Forest	19.193	84.644	26.737
Extra Tree	15.747	68.678	22.630
AdaBoost	22.955	76.651	29.370
Gradient Boost	14.709	91.526	26.357
XGBoost	15.132	90.575	26.495
CatBoost	15.617	67.597	22.465

A machine-learning model was developed[22] which could predict the mechanical properties of stir-cast hybrid metal-matrix composites. Experimental data on such properties were compiled, and used to train various machine-learning models. The data-set was divided groups termed training and testing, in ratios of 70:30, 80:20 or 90:10. For each ratio, all of the machine-learning regression models were applied to the training data-set and each trained model was checked using the testing data-set. The predicted output values of the testing data-set for each model were compared with the actual UTS values, and R^2 values were calculated (tables 3 and 4). For every regression model, the R^2 score for the 80:20 split gave better results when compared with other ratios. The decision-tree model (80:20 ratio) gave a maximum accuracy of 92.029%, and was lowest for AdaBoost regression, with an accuracy of 82.061%. A decision-tree regression model offered the greatest accuracy in the prediction of the ultimate tensile strength of such composites (figure 8). The ultimate tensile strengths of two composites, having AA7075 or AA6061 as the matrix, were predicted using the decision-tree model. Comparison with experimental data revealed an error of less than 10%. The UTS of samples of composites having AA7075 and AA6061 as the matrix were predicted using the decision-tree model and compared with actual experimental values. The error for the Al7075 hybrid composite was 5.533% and that for the Al6061 hybrid composite was 9.903%.

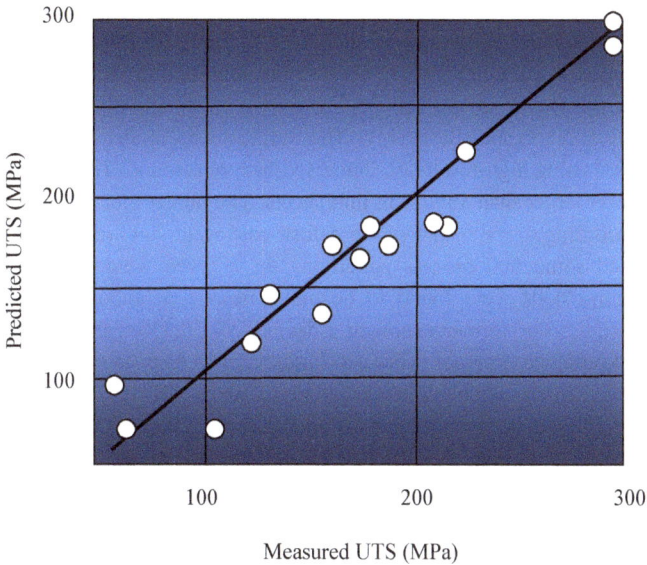

Figure 8. Predicted versus actual UTS using decision tree regression

The atmospheric corrosion of aluminium alloy was predicted[23] by using a back-propagation neural network. The artificial neural network model comprised 7 input nodes, 5 hidden layer nodes and 1 output node, and the correlation coefficient for the model was 0.8821. The predicted results were close to the experimental data.

Micro-arc oxidation of AA2024 was used[24] to improve the anti-corrosion behaviour. The corrosion resistance was evaluated by salt spray testing. A parallel approach combined electrochemical characterization with artificial intelligence to predict corrosion. The method generated up to 50 parameters that were treated by an artificial neural network. Following training, the average accuracy of the predicted corrosion resistance after 200h was up to 99%.

The evaluation of various machine-learning approaches, including principal component analysis, k-nearest neighbour, multilayer perceptron, single vector machine and random forest, as applied to friction stir welding was reviewed[25]. The input variables were divided into two groups, with one group comprising application variables and the other group comprising friction stir welding variables. The application variables covered the aluminium alloy, the joint configuration, the sheet thicknesses, the initial mechanical properties and the chemical composition. The friction stir welding variables covered the rotational speed, the travel speed, the forging force, the longitudinal and transverse forces, the torque and the specific energy. The output to be modelled was the defect index, as quantified using high-resolution immersed-bath ultrasound. The latter technique could detect defects larger than 150µm in thin sheets. The defect index was in turn spilt into 5 classes, according to the nature of the defect: cold weld, hot weld and width. The data-set comprised some 500 process conditions for a given weld regime. The best machine-learning methods were found to be the K-nearest neighbour and multilayer perceptron algorithms. The former exhibited a deviation of 0.55 on the defect index as compared with experimental values. This was slightly better than that of the multilayer perceptron model, which exhibited a deviation of 0.69. Of the initial 59 model parameters, 10 to 15 of them were retained in the final algorithms. The main predictors included the material thickness, the ultimate tensile strength of the base material, the rotational speed, the travel speed, the weld forces and the specific energy. The K-nearest neighbour model provided a map of defect indices with regard to rotational speed and travel speed, but this was possible only when a higher density of data was available.

Continuous dynamic recrystallization, occurring during the friction stir welding of AA6082-T6 aluminium alloys, was modelled[26] by linking a properly trained neural network to a finite-element model of the process. The network employed used, as inputs, local values of the strain, strain-rate and temperature. It was trained by starting with experimental data and numerical results in order to predict the average grain size. The complex geometry of some joints makes it difficult to propose overall governing equations for the theoretical analysis of friction stir-welding. The required parameters were determined[27] by using artificial neural networks. In order to train the latter, experimental test results on thirty AA7075-T6 specimens were considered, and analyzed using a back-propagation algorithm (figure 9). The network-testing was carried out by using experimental data that were not used during network-training. The rotational speed of the tool, the welding-speed, the axial force, the shoulder-diameter, the pin-diameter and the tool hardness were used as inputs to the networks. The yield strength, the tensile strength, the notch tensile strength and the hardness of the weld-zone were the outputs of the neural networks (figures 10 to 13). The predicted hardness of the weld-zone, the yield strength, the tensile strength and the notch-tensile strength exhibited, in order, the lowest mean relative error (table 5).

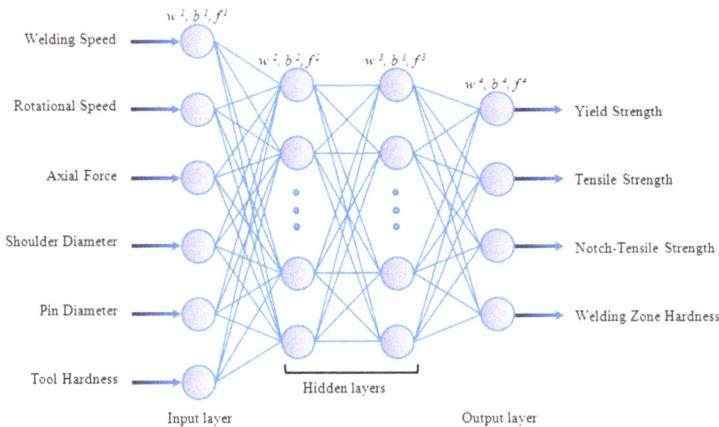

Figure 9. Conceptual structure of the artificial neural network: 4 layers with full interconnection. The 6 input parameters are logged into the input layer to determine 4 outputs. Reproduced from: Artificial neural networks application for modeling of friction stir welding effects on mechanical properties of 7075-T6 aluminium alloy, by E. Maleki, IOP Conference Series: Materials Science and Engineering, 103[1] 2015, 012034 under Creative Commons Licence.

In a further study Al7075, with and without aging, was joined[28] by friction stir welding and tensile-tested. The results were analyzed using neural networks. The tensile strength of the assemblies was again estimated by using, as input parameters, the type of material, the number of revolutions and the feed-rate. In the neural network model, various nerve numbers, divers transfer functions and varied layer-numbers were essayed. The prediction-performance was judged by using neuron and algorithm-type variables. The ANN model for parts joined using friction stir welding gave predictions of high consistency. The best estimates were obtained when using the Levenberg-Marquardt algorithm and 3 neurons, leading to a 98% success-rate. Artificial intelligence was again used[29] to model the friction stir welding of the dissimilar AA5083-O and AA6063-T6 aluminium alloys. The 27 experiments were planned using a so-called L_{27} orthogonal 'Latin-square' type of array (table 6). The experimental results were then used to develop an artificial neural network mathematical model for the tensile strength, microhardness and grain size. A hybrid approach, involving the artificial neural network and a genetic algorithm, was used for multi-objective optimization. The models for tensile strength, microhardness and grain size were reliable, with average percentage prediction errors of 0.053714, 0.182092 and 0.006283%, respectively. Selecting for the best fitness value, 50 so-called Pareto optimal solutions were obtained which offered various solutions corresponding to divers process parameters. In Bayesian multi-objective optimization an expected hypervolume improvement is often used to measure the goodness of candidate solutions but, when there are many objectives, the calculation of that improvement can be computationally prohibitive. An alternative approach is to measure the goodness of a candidate solution based upon the distance of that candidate from the Pareto front in objective space. This Bayesian many-objective optimization algorithm has been applied[30] to a hyper-parameter selection problem and to high-temperature creep-resistant alloy design.

Table 5. An L_{27} experimental matrix for the friction stir welding of AA5083-O and AA6063-T6

Test	TRS(r/min)	WS(mm/min)	SD(mm)	PD(mm)	TS(MPa)	MH(H$_v$)	GS(μm)
1	700	40	15	4.5	136.2	59.53	19.886
2	700	40	18	5.0	146.3	69.84	15.625
3	700	40	21	5.5	141.1	64.03	16.203
4	700	60	15	5.0	145	66.74	16.509
5	700	60	18	5.5	150.2	73.4	10.937
6	700	60	21	4.5	143.7	65.88	16.826
7	700	80	15	5.5	135.1	54.14	19.021
8	700	80	18	4.5	138.8	62.09	18.657
9	700	80	21	5.0	139.6	62.53	18.121

10	900	40	15	5.0	148.8	69.23	14.583
11	900	40	18	5.5	153.5	76.41	11.82
12	900	40	21	4.5	152.6	68.77	14.112
13	900	60	15	5.5	150.9	75.54	11.217
14	900	60	18	4.5	161.2	85.25	8.578
15	900	60	21	5.0	156.1	78.56	9.943
16	900	80	15	4.5	146.3	62.71	18.229
17	900	80	18	5.0	151.3	72.87	12.323
18	900	80	21	5.5	145.2	65.14	18.229
19	1100	40	15	5.5	145.4	69.92	13.257
20	1100	40	18	4.5	151.2	73.22	12.86
21	1100	40	21	5.0	143.9	72.43	12.152
22	1100	60	15	4.5	150.1	71.21	13.67
23	1100	60	18	5.0	157.5	79.39	9.72
24	1100	60	21	5.5	152.5	76.36	11.513
25	1100	80	15	5.0	137.5	66.54	17.156
26	1100	80	18	5.5	147.5	72.15	12.152
27	1100	80	21	4.5	142.5	62.45	17.5

TRS: tool rotational speed, WS: welding speed, SD: shoulder diameter, PD: pin diameter, TS: tensile strength, MH: microhardness, GS: grain size

In the case of the genetic algorithm-based optimization, several solutions were obtained instead of a single optimum solution. The latter was defined to be the solution which satisfied the manufacturing aims, and this in turn depended upon the values of each response. Any particular optimum solution could be selected on the basis of the improvement in quality for a given set of process parameters: a solution yielding the maximum tensile strength and microhardness and the minimum grain size. The optimum set of process parameters consisted of a tool rotational speed of 881.9627r/min, a welding speed of 61.59241mm/min, a shoulder diameter of 16.838mm and a pin diameter of 4.966mm.

Artificial neural networks were used[31] to predict the quality of the friction stir welded surface of AA6082-T6 sheets. Nine different variables were recorded during the welding. These were the forces and accelerations in three spatial directions, the spindle torque, and the temperatures at the tool shoulder and tool probes. In one case, the welds were classed as good or defective on the basis of visual inspection of the weld surface. In another case, the welds were separated into two classes on the basis of surface topography analysis. Three different artificial neural network architectures were tested with regard to their ability to predict surface quality. These were termed: feed forward fully connected, recurrent and convolutional. The highest accuracy was found when the convolutional neural network was used. Evaluation of the force which was transverse to the welding direction furnished the highest accuracy (99.1%) for the prediction of the result of visual inspection. The accuracy of the prediction of the topographic analysis was 87.4% when the spindle torque was evaluated. By using all nine of the process variables to predict the topographic analysis, the accuracy could be improved to 88.0%.

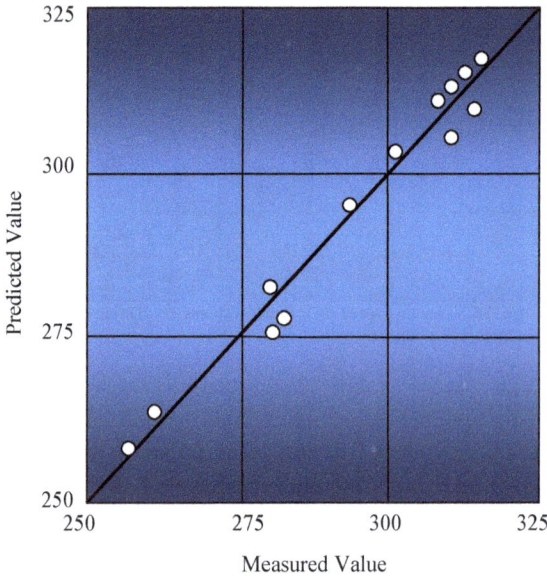

*Figure 10. Predicted versus experimental values of the yield strength of
15 samples of friction stir-welded AA7075-T6 aluminium alloy*

An important aspect of the friction stir welding of parts is the differing hardness levels which are found, in the characteristic welding zone, as a function of the maximum temperature reached during welding. Such differences affect the mechanical properties and service life of the component. A hybrid model was developed[32] for predicting the final hardness of individual points of the weld as a function of the maximum temperature attained. The hybrid approach combined the finite element method and an artificial neural network. A finite-element model temperature map was introduced into the neural network, together with experimental results, for the network training. This approach provided a reliable framework for forecasting the hardness, following friction stir welding, without having to investigate each weld.

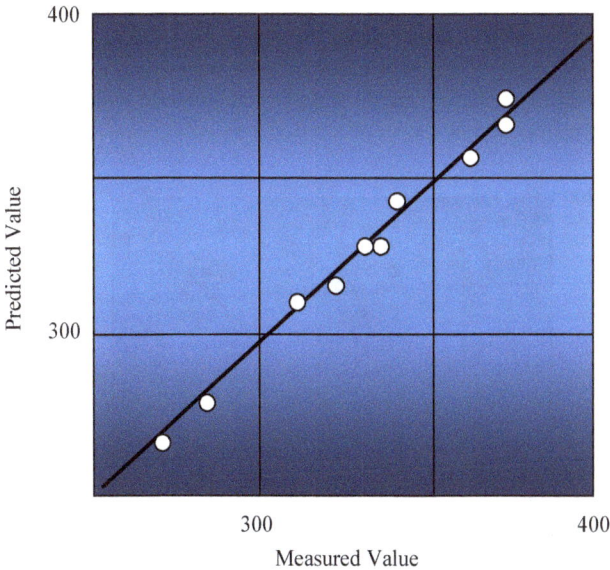

Figure 11. Predicted versus experimental values of the tensile strength of 15 samples of friction stir-welded AA7075-T6 aluminium alloy

Unsuitable settings of the parameters can be detected by examining surface defects such as excess flash formation or surface galling. Two different learning-based algorithms were applied[33] to the improvement of the surface topography of the welds. The surfaces of 262 welds were evaluated for the purpose of reinforcement learning, and Bayesian optimization was used to determine the most suitable welding speed and tool rotational speed. The optimization problem was solved by using reinforcement learning via value iteration. The latter algorithm was inefficient because all of the actions and states had to be iterated over, and each possible parameter-combination had to be evaluated in order to identify the best policy. It was better to solve the optimization problem directly by using Bayesian optimization. There was a choice as to whether or not information from other studies was used. Both Bayesian optimization approaches identified suitable welding parameters much faster than did a random search algorithm. A data-driven method was used[34] to forecast the mechanical behaviour of AA6061-T6 with regard to friction stir welding. The gated recurrent unit concept, a deep learning model, was used for the first time to forecast stress-strain curves. The technique could model time-series data and relied only upon historical and actual data concerning the investigated material. The performance of the technique was demonstrated by using data which were collected from uniaxial tensile testing of the base material and of friction-stirred welded material; both tested at a deformation rate of 0.001/s.

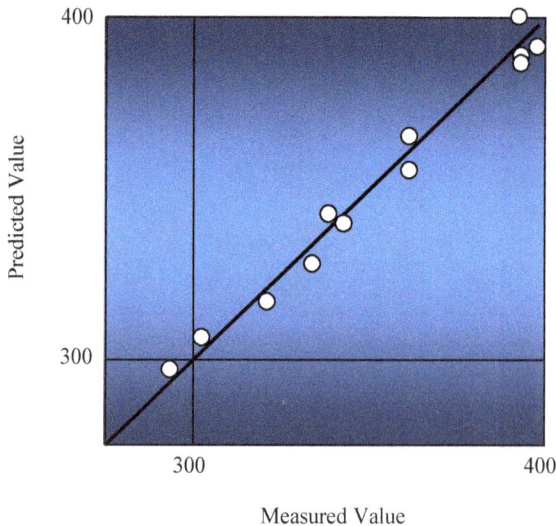

Figure 12. Predicted versus experimental values of the notch tensile strength of 15 samples of friction stir-welded AA7075-T6 aluminium alloy

Supervised machine learning regression and supervised machine learning classification models have been used[35] to predict the ultimate tensile strength and weld-joint efficiency of friction stir welded joints. A polynomial regression model gave better results than did other supervised machine learning regression models. Decision tree (Gini index or information gain criterion) and artificial neural network classification models furnished better classification results than did k-nearest neighbour classification models.

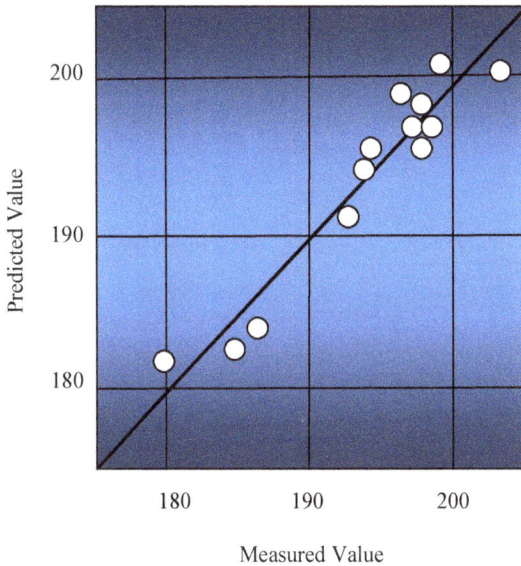

Figure 13. Predicted versus experimental values of the weld-zone hardness of 15 samples of friction stir-welded AA7075-T6 aluminium alloy

A neurobiological-based unsupervised machine-learning algorithm, a self-organizing map neural network, was used[36] to predict the fracture location in dissimilar friction stir welded AA5754 and C11000 alloys. The tool shoulder diameter, the tool rotational speed and the tool traverse speed were the input parameters, while the fracture location was the output. The location could be in the thermomechanically affected zone of the copper alloy or in the equivalent location of the aluminium alloy. Among competing algorithms, the self-organising map algorithm was able to predict the fracture location with the highest accuracy (96.92%).

A dissimilar copper/aluminium lap-joint was generated[37] by means of force-controlled hybrid friction diffusion bonding. During welding, the torque, the welding force and the plunge-depth were recorded. Due to the force-controlled process, to tool wear and to the use of differing materials, the data series varied significantly and impaired quality assurance. A convolutional neural network was developed which permitted evaluation of the processing data. Data from sound welds as well as from samples with weld defects were included. Deviations from ideal conditions, due to tool wear and to the use of differing alloys, were considered. The validity of the approach was determined by cross-validation during training. With its **88.5%** accuracy, the convolutional neural network was a suitable tool for monitoring the processes.

Table 6. Accuracy of property predictions

Property	PCC	MRE(%)
yield strength	0.99543	1.0820
tensile strength	0.99875	1.1303
notch tensile strength	0.99403	1.3046
weld-zone hardness	0.99918	0.7917

PCC: Pearson correlation coefficient, MRE: mean relative error

Ultrasonic stationary shoulder assisted friction-stir welding is a promising means for joining aluminium- and magnesium-based alloys. Pigeon-inspired optimization, a swarm intelligent optimization algorithm, was proposed for the mathematical modeling and processing-optimization of the technique. A pigeon-inspired optimized artificial neural network was used to identify relationships between the inputs and output of the welding process. A reliable model was developed, and a joint with a tensile strength of 161MPa was produced; a strength higher than that ever reported for similar welding[38].

The rotation-velocity, welding-speed, zinc-interlayer thickness and ultrasound-power were optimized by using a back-propagation neural network and grey-wolf optimization algorithm so as to achieve the high-quality Zn-added ultrasound-assisted friction-stir lap-welding of aluminium alloy and AZ31B magnesium alloys[39]. The predictive accuracy of the trained back-propagation neural network model was clearly acceptable and the optimum processing parameters were a rotation velocity of 1054rpm, a welding-speed of 54mm/min when the zinc-interlayer thickness was 0.05mm and the ultrasound-power was 1568W. The tensile shear load of the joint attained 9.05kN and the strength was 11.8% greater than that of previous optimum joints. In similar work[40] a hybrid radial bias function neural network and genetic algorithm were used to optimize the processing parameters for the joining of AA7075-T6 and AZ31B. The optimum conditions in this case were a rotation-velocity of 1037rev/min, a welding-speed of 82mm/min, a zinc interlayer-thickness of 0.04mm and an ultrasound-power of 1440W. The tensile shear load of the joint was 8.860kN.

Biological evolution concepts were used to create new and effective bio-inspired computer optimization algorithms. Nine specimens of AA6262 were friction stir welded, with the tool rotational speed, traverse speed and plunge depth being used as input parameters, and the ultimate tensile strength as the output parameter[41]. The aim was to deduce the maximum optimized ultimate tensile strength by using bio-inspired artificial intelligence algorithms such as differential-evolution and Max Lipschitz optimization. The results showed that the differential-evolution algorithm predicted a slightly higher value of the ultimate tensile strength (184.87MPa) as compared with Max Lipschitz optimization, which gave an ultimate tensile strength value of 183.94MPa.

The same problem afflicts resistance spot-welding; that is, an inconsistent quality of the welds. The important parameters here are the welding current, the electrode force and the welding time. In addition, the electrical resistance of the aluminium alloy depends upon the thickness of the material. Parameter prediction was achieved[42] by using an artificial neural network. The network was tested for predicted weld-quality by correlating the input parameters and the tensile shear strength of the aluminium alloy. The results of tensile shear strength testing and parameter optimization were applied to the welding process. The resultant accuracy was a mean-squared error and accuracy of 0.054 and 95%, respectively. The resistance spot welding of 2219/5A06 aluminium alloy was analysed[43] using artificial intelligence methods in order to improve the efficiency of weld-quality evaluation. A multi-signal fusion method combined principal component and correlation analyses. A back-propagation neural network model was optimized by using a sparrow search algorithm, with the inputs and outputs of the model being the variables following multi-signal fusion and the button diameter, respectively. The sparrow search algorithm is a bionic intelligent optimization algorithm, which provides the advantages of reliable optimization ability and rapid convergence. It is best described as being an explorer–follower–warner model. As compared with the standard back-propagation neural network model, the sparrow model reduced the mean absolute error by 42.33%, the mean square error by 51.84% and the root mean square error by 31.45%. Its R^2 coefficient attained 0.6482; much higher than that (0.2464) for back-propagation. According to the above indicators, the performance of the sparrow model was better than that of the standard back-propagation neural network model. When compared with other models, the evaluation performance of the sparrow model was best (table 7), and it could successfully predict abnormal spot welds. The standard sparrow algorithm randomly generates individual location information and thus the resultant target is not optimum and this affects iterative efficiency and the error rate. The sine map has a simple mathematical form and offers convenient parameter adjustment. It was therefore used to improve the sparrow algorithm. The sequences which were generated then had a more chaotic nature.

Table 7. Evaluation indicators for various prediction models

Prediction Model	MAE	MSE	RMSE	R^2
BP	0.9805	1.6618	1.272	0.2464
GA-BP	0.8336	1.172	1.077	0.4762
PSO-BP	0.8011	1.0517	1.0194	0.5402
SSA-BP	0.6869	0.8099	0.8908	0.6458
Sine-PSO-BP	0.7754	1.0237	0.991	0.5728
Sine-SSA-BP	0.6889	0.7763	0.8719	0.6482

BP: back-propagation, PSO: particle swarm optimization, SSA: sparrow search algorithm, MAE: mean absolute error, MSE: mean square error, RMSE: root mean square error

Neural networks were used[44] to predict the solidus temperatures of hypoeutectic Al-Si-Cu casting alloys when cooled at various rates. Knowledge of the solidus temperature permitted the choice of melt-treatment and casting temperature. At the time, there was only one reported equation for the determination of the solidus temperature for hypoeutectic aluminium alloys. A computational algorithm was developed which compared various models for predicting the solidus temperature, with regard to the influence of alloying elements. There was a good correlation between experimental and calculated results and a neural network was concluded to possess great potential for the modelling of the solidus temperature of Al-Si-Cu alloys.

The hot-rolling of AA5083 alloy was simulated[45] by using finite-element methods. The temperature distribution within the roll and the slab, and the stress and strain-rate fields, were deduced from a steady-state analysis of the process. It was assumed that the thermo-viscoplastic behaviour of the material was described by the Perzyna constitutive equation, and that the rolling occurred under plane-deformation conditions. The geometry of the slab, the load, the rolling speed, the percentage thickness reduction, the initial thickness of the slab and the friction coefficient were expressed in parametric form. The convergence of the results was evaluated by using available experimental and theoretical data. Because finite-element simulation of the process was time-consuming, an artificial neural network was based upon the back-propagation method. The output of the finite-element simulation was used to train the network, and the latter was then used to predict the behaviour of the slab during hot-rolling.

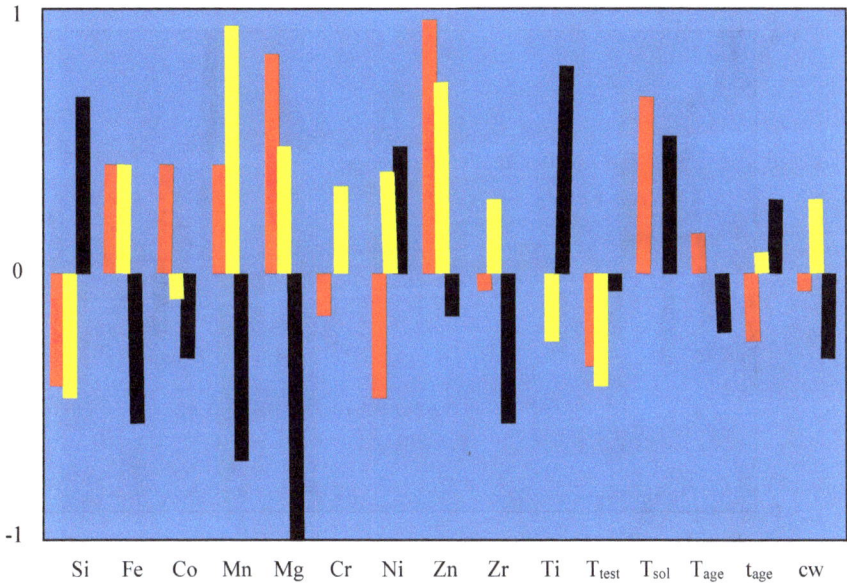

Figure 14. Sensitivity plots of trained ANN models for the yield stress, ultimate tensile strength and elongation, using the connection weight method for low temperatures, red: yield stress, yellow: UTS, black: elongation, T_{test}: test temperature, T_{sol}: solutionizing temperature, T_{age}: aging temperature, t_{age}: aging temperature, cw: cold work.

A model based upon least-square support vector machines was used for the first time[46] to predict the mechanical and electrical properties of Al-Zn-Mg-Cu alloys. Data-mining and artificial intelligence techniques were used to check the forecasting capability of the model. In order to improve the predictive accuracy and generality of the model, grid algorithm and cross-validation techniques were used to determine automatically the optimum hyper-parameters of the model. The present model offered slightly better generalized predictions than did a back-propagation network combined with a gradient descent training algorithm. With its advantages of computational speed, unique optimum solution and generalization capability, the present model was considered to be useful as an alternative tool for the aging-optimization of aluminium alloys. A non-dominated sorting-based multi-objective genetic algorithm was further proposed for making trade-offs between the mechanical and electrical properties.

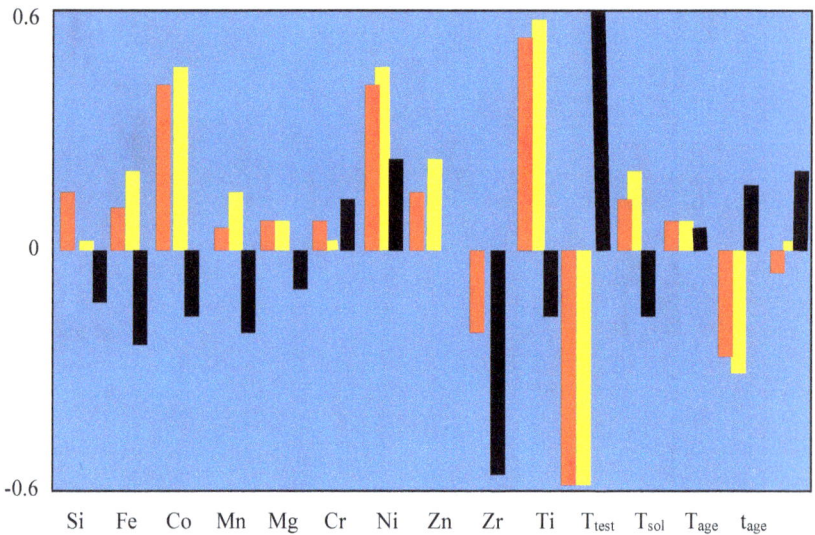

Figure 15. Sensitivity plots of trained ANN models for the yield stress, ultimate tensile strength and elongation, using the connection weight method for high temperatures , red: yield stress, yellow: UTS, black: elongation, T_{test}: test temperature, T_{sol}: solutionizing temperature, T_{age}: aging temperature, t_{age}: aging temperature, cw: cold work.

An approach which was based upon genetic algorithms, swarm-intelligence optimization and finite-element methods was used[47] to model the processing of aluminium-matrix nanocomposites. Ceramic nanoparticles were added to the aluminium alloy in order to investigate experimentally the microstructures and mechanical behaviours of metal-matrix nanocomposites. Inspired by the idea of breeding-swarms, a genetic algorithm and particle-swarm optimization (PSO) hybrid was used to combine the standard velocity and position up-dating rules of PSO with the concepts of selection, crossover and mutation, as borrowed from genetic algorithms. Experimental results confirmed the efficiency of the proposed model for estimating the optimum processing conditions for preparing nanocomposite by casting.

Computational intelligence-based methods have been used[48] to design novel age-hardenable aluminium alloys by exploiting the effects of all of the precipitate-forming elements taken together and crossing the limits of the compositions defined for the various series. A data-base was created from the tensile properties of age-hardenable aluminium alloys in the 2XXX, 6XXX and 7XXX series. The data were classified by testing temperature and various models were developed for the tensile properties within the different temperature regimes by using artificial neural networks. The inherent relationships between the composition, processing variables and mechanical properties were explored via sensitivity analysis. It is important to possess a clear understanding of the relative effects of the alloying additions upon the properties of the alloys. It is difficult to identify the effect of the input variables upon the output of a system, in case of ANN models, due to complex hidden interactions. Sensitivity analysis automatically identifies all of the relevant parameters among a set of potential parameters. The sensitivity analysis is performed on a trained neural network. Such an analysis was performed for the yield strength, ultimate tensile strength and elongation, using a connection-weight method for low-temperature, room-temperature and high-temperature data respectively. The sensitivity plot for low temperatures showed that most of the alloying elements, apart from silicon and nickel, had a positive effect upon the strength and a negative effect upon the ductility. Additions to aluminium alloys promoted strengthening by solid-solution hardening or precipitation-hardening. Both processes had a negative effect upon the ductility. The sensitivity plot for room temperature (figure 14) also showed that most alloying elements had a positive effect upon strength. Silicon had an anomalous positive effect upon ductility. The sensitivity plots exhibited similar trends, but were different for the high-temperature regime (figure 15): the ageing precipitates could not resist softening, due to their coarsening. Intermetallic compounds such as Al_3Ti could here act as strong barriers to dislocation motion. This was reflected in the strengthening effect of titanium. On the other hand, no such effect was exhibited by zirconium even though it forms an analogous compound with aluminium. In all cases, the improvements in strength and ductility tended to depend upon different variables. In order to design alloys satisfying the conflicting objectives of high strength and good ductility, multi-objective genetic algorithms were used to search out optimum solutions, with the ANN models acting as objective functions. A tailored composition, which went beyond any of the then-existing age-hardenable series, was developed experimentally.

Table 8. Rule base for the yield strength of AA2XXX alloys

Si	Cu	Mg	Zn	T_{aging}	t_{aging}	Cold Work	Yield Strength
high	medium	high	high	high	low	low	low
medium	medium	medium	high	low	high	low	low-medium
low	medium	high	high	high	low	high	medium
low	high	low	low	low	high	high	high-medium
low	high	low	low	high	medium	low	high

An artificial intelligence based computational approach was used[49] to improve both the ductility and the strength of age-hardenable AA2XXX, AA6XXX and AA7XXX alloys by modifying the composition and heat-treatment parameters. Published data were used to develop the models for tensile strength, yield strength and elongation. Clustering was used to gather the variables in the database into various levels and generate a fuzzy rule which correlated those variables. An adaptive neuro-fuzzy inference system used the fuzzy rules to develop data-driven fuzzy predictive models for the properties of the alloys. The models in turn played the role of objective functions for multi-objective optimization, using a genetic algorithm to handle the conflicting objectives of improving ductility as well as strength. The rules in a fuzzy set are if-then statements that provide conditional factors of fuzzy logic. The fuzzy rules (tables 8 to 10) which were used for the yield strength assumed that, if x is A and y is B then z is C. A clustering method was used to develop fuzzy clusters of the data for each variable. The fuzzy clusters were then customized by naming them low, medium, high, low-medium and high-medium. There were 5 rules for the yield strength of AA2XXX and AA7XXX aluminium alloys series and 7 rules for the yield strength of AA6XXX alloys. The optimum solution improved with increasing numbers of rules. The numbers of rules for the ultimate tensile strength of

AA2XXX, AA6XXX and AA7XXX aluminium alloys were 5, 7 and 5, respectively. In the case of the elongation, the rules for the AA2XXX and AA7XXX series were 5 in number while, in the case of AA6XXX alloys, 12 rules were used. One hundred non-dominated (Pareto) optimum solutions were generated for the 3 series by multi-objective optimization. In the case of AA2XXX alloys, copper and magnesium had the greatest influence together with, to some extent, silicon. Silicon, magnesium and copper were also important for the AA6XXX alloys. In the case of AA7XXX alloys, only zinc and magnesium were important. In this alloy series, the amounts of those elements were almost the same for all non-dominated solutions. Elements other than those singled-out above also made some contribution; especially to the AA6XXX and AA7XXX alloys.

Table 9. Rule base for the yield strength of AA6XXX alloys

Si	Cu	Zn	Mg	T_{aging}	t_{aging}	Cold Work	Yield Strength
low-med.	low	medium	low	high	low	low	medium
medium	medium	high	low	high	low	low	high-med.
medium	medium	high	low	low	high	low	high-med.
low	low-med.	low	low-med.	low	high	low	high
low-med.	low	medium	medium	medium	low	high	low
high	high	high	high	high	high	low	low-med.
high	high	high	high	low	low	low	low-med.

The grain-size and its measurement are pivotal factors in preparing high-strength aluminium-alloy components. This is now aided by digital image-processing and pattern-recognition technology as applied to quantitative metallography. A digital image-processing method for determining the grain size from metallographic images was developed[50]. In order to determine the grain-size from digital metallographic images, digital processing was used to identify grain boundaries by employing a novel edge-detection algorithm which was based upon fuzzy logic. Numerous metallographic images of various qualities were used to validate the method in accord with ASTM standards.

Table 10. Rule base for the yield strength of AA7XXX alloys

Si	Cu	Mg	Zn	T_{aging}	t_{aging}	Yield Strength
low	medium	low	medium	medium	high	low
high	high	high-medium	high	low	low	low-medium
medium	low	high	low	low	low	medium
high	medium	medium	medium	high	high	high-medium
low	medium	low	medium	medium	low	high

Another in-service behaviour which must already be considered during the alloy-design stage is wear. In this regard the dry sliding wear of a hybrid aluminium-matrix composite has been evaluated[51] for Al6061 containing 3 to 15wt% of Si_3N_4 and nanographite powder. The two additions, 50wt% of each, were blended in a high-energy ball-mill and homogeneously mixed to produce a sound composite by stir-casting. The wear-rate was measured against an EN32 steel disc, using a pin-on-disc tribometer. An integrated response-surface method and a genetic algorithm were used to optimize the pin-on-disc process-parameters. Analysis of the variance showed that the sliding-distance played a major role in the dry sliding wear-rate, followed by the load, the sliding speed and the additions used … with two-factor interactions and quadratic terms also making significant contributions. The genetic algorithm indicated a minimum wear-rate of 0.827mg under optimum conditions. Scanning electron microscopy revealed that very fine grooves were present under optimum conditions whereas, for other settings, severe ploughing occurred. With increasing speed, and therefore with an increase in temperature at the rubbing surfaces, there was a change in wear mechanism from abrasive to adhesive.

The use of neural networks in predicting the wear loss of aluminium-copper silicon carbide composites was examined[52]. The effects of adding copper as an alloying element and silicon carbide as reinforcement particles, to a Al-4wt%Mg metal matrix, were investigated. Various Al-Cu alloys and composites were subjected to dry sliding wear, using pin-on-disk apparatus, under a 40N normal load and a rotational speed of the counter-face disk of 150rpm at ~20C (~50% relative humidity). Experimental data were coded before training a feed-forward back-propagation artificial neural network. The predictions were compared with experimental results. The average value of the absolute relative error of non-coded values reached 2.40%.

Artificial intelligence tools were used[53] to predict the evolution of crater tool wear. Tests were performed which involved using tungsten carbide cutting tools to machine AA7075 on a CNC lathe. The corner-radius, feed-rate, cutting-speeds and cut-depth were studied with regard to tool crater wear. Of 30 experiments, 24 were used for model-training and 6 for testing. Another experiment was performed under different cutting conditions in order to check the models. By employing a hyperparameter search and careful tuning, the optimum learning-rate for each model was determined in order to ensure effective convergence. It was concluded that the gradient boost model was the best, as judged by an R^2 of 0.9085, an MAE of 0.05425, an RMSE of 0.06635, an RAE of 0.24265 and an RSE of 0.09115; with deviations between predicted and measured crater tool wear of 8.27%.

A fuzzy-logic artificial intelligence technique was used[54] to predict the machining performance of Al-Si-Cu-Fe die-casting alloy which was treated with additives such as strontium, bismuth and antimony in order to improve the surface roughness. A Pareto-ANOVA optimization method was used to determine the optimum settings for machining. Experiments were performed using oblique dry-turning. The machining parameters of the cutting-speed, the feed-rate and the depth-of-cut were optimized with respect to the surface roughness. A cutting-speed of 250m/min, a feed-rate of 0.05mm/rev and a depth-of-cut of 0.15mm were the optimum dry-turning settings. Strontium and antimony had detrimental effects upon workpiece machinability. Samples which contained bismuth exhibited the lowest surface-roughness values and this was attributed to the presence of pure bismuth, which acted as a lubricant during turning. A check of the validity of the model indicated a surface-roughness error-rate of only 5.4%.

A study was made of the improvement of the surface roughness parameters for milled AA6061 alloy using carbide cutting tools coated with chemical vapour deposited TiCN under dry conditions[55]. A model estimating the surface roughness used artificial neural networks and response surface methodology. The cutting-speed, depth-of-cut and feed-rate were the input parameters for experimental design. For neural-network modelling, a back-propagation algorithm was used for the optimum selection of data for training. Five learning algorithms were used: conjugate gradient back-propagation, Levenberg-Marquardt, scaled conjugate gradient, quasi-Newton back-propagation and resilient back-propagation. Analysis-of-variance showed that the depth-of-cut was the most effective parameter affecting surface roughness. The data estimated using neural networks and response surface methods were close to the experimental data.

An experimental investigation was made[56] of the modelling and optimization of the heat-affected zone in pulsed Nd:YAG laser-cut Duralumin sheet. The quality was improved by a proper control of process parameters such as the gas-pressure, the pulse-width, the pulse-frequency and the scanning-speed. Optimization of the heat-affected zone was carried out by using the hybrid approach of multiple regression analysis and genetic algorithms. In this method, a second-order regression model was developed by using multiple regression analysis with the help of experimental data obtained using an L_{27} orthogonal array.

A cellular automaton method, assisted by a back-propagation neural network, was proposed[57] for the simulation of grain and pore growth. The technique first used the neural network to identify relationships between the porosity and the solidification parameters, and then used those relationships to deduce further transformation rules for pore growth, as they applied to A356 alloy. The shapes and volume fractions of pores, as observed experimentally were consistent with the predictions. The method could reduce the difficulty of simulating the entire casting process without having to solve the high-dimensional equations which governed porosity.

Machine-learning models were proposed[58] for porosity detection, in microstructural images of wire-arc additively manufactured AA6061 parts, using limited data-sets. The distinguishing of pores from microstructures was achieved using pixel-level colour and textural features identified by using Gabor filters. The machine-learning models, in which hyperparameters were chosen by cross-validation, produced an average classification accuracy of 98.89% (random forest) for the detection of pores larger than 5μm. Experimental results showed that the methods were very effective.

Combined artificial intelligence and plasma emission spectroscopy techniques were used[59] to identify the porosity of additively manufactured parts during the process. Time- and position-synchronized spectra were collected during the directed-energy deposition manufacture of an AA7075 part. Eighteen features, extracted from the spectra, were correlated with deposition parameters as gleaned by 3-dimensional X-ray computed tomography scanning. These were used to train a random forest classifier. The well-trained classifier then attained up to 83% precision in the porosity recognition of deposits. The feature-importance, as judged by the random forest classifier indicated that the intensities of spectra at wavelengths of 414.234nm and 396.054nm, and the kurtosis of spectra over wavelength-ranges of 484 to 490nm and 508 to 518nm were the most effective features for recognizing porosity.

A machine-learning framework was proposed[60] for the prediction of changes in the local strain distribution, plastic anisotropy and failure during the tensile testing of AlSi10Mg alloy samples produced by means of selective laser melting. The framework combined aspects of additive manufacturing and artificial intelligence and included the printing of specimens via laser powder-bed fusion and the use of X-ray computer tomography to measure the internal defect-distribution. Mechanical testing and digital image correlation monitored the local strain evolution and the extracted data drove the training and evaluation of an artificial neural network model. Features such as the size, shape, volume fraction and distribution of porosity were used as input to the neural network. The latter model successfully predicted the evolution of local strains, plastic anisotropy and failure during tensile deformation. The intensity and location of strain concentrations, as well as the shape of shear bands and the locations of crack initiation were well-predicted.

A machine-learning method was used to establish the relationship between alloy composition, thermomechanical processing and the forming limit diagrams of AA5XXX and AA6XXX alloys[61]. The thermomechanical processing parameters included homogenization, and hot or cold rolling, together with mechanical properties. A 2-stage machine-learning model was trained using this database. In the first stage the minimum and maximum points of the minor strain were predicted by using a data-set which included chemical compositions and processing parameters and which was analysed by support vector regression. In the second stage the predicted minor strain from the first stage was used as an input, in addition to the same data-set, to predict the major strain by using gradient boost regression. The trained machine-learning model successfully predicted forming limit diagrams with an R^2 better than 0.93. As a practical check, finite-element simulations were made of a cross-die were performed. The difference in draw depth of the cross-die, for predicted and measured forming limit diagrams, was between 5 to 10%.

A pattern-recognition based method was proposed[62] for determining the forming limit curve in sheet-metal forming. The curve is defined by the onset of necking, is based upon the cross-section and assumes that failure occurred due to clear localized necking. This approach fails in the case of brittle materials, which do not exhibit distinct necking. A first approach was based upon crack evaluation, which is well-defined, but it is also important to evaluate the general material instabilities that precede failure. An analysis of the material behaviour during stretching was carried out in order to characterize various classes of instability. The results of Nakajima tests were studied by using an optical measurement system. A conventional pattern-recognition method, based upon textural features, was applied to the determination of the forming limit curve. The results revealed accurate prediction of the onset of necking, even in the case of high-strength materials. Deviations were noted in the determination of diffuse necking. The method was later applied[63] to the temporal and spatial determination of the onset of local necking, as determined by means of a Nakajima test, for a DC04 deep drawing and a DP800 dual-phase steel as well as for an AA6014 aluminium alloy. Attention was focused on the progress of the necking process and its transformation throughout the remainder of the forming. The strain behaviour was monitored using a machine-learning approach, on the basis of images recorded when the material was close to failure. The learned failure-characteristics were then transferred to other forming sequences so that critical features which indicated failure could be identified at an early stage and thus permit the determination of the onset of necking. The growth behaviour and the traceability of the necking area were objectified by the weakly-supervised machine-learning approach. New optimization algorithms, involving neural networks, were used[64] to maximize the formability of sheet metal on the basis of the tensile curve and texture of aluminium sheet. Experimental and numerical studies were made of the effect of texture and tensile properties. Texture-effect evaluation was carried out by using Taylor homogenization. The resultant data were used to train an artificial neural network by means of various optimization methods. These included the grey wolf algorithm, and chimp optimization algorithm and the whale optimization algorithm. The results showed that, for aluminium alloys, the preferred texture for formability is the cube texture. Slight differences in the tensile behaviour of aluminium sheets under similar conditions caused no marked decreases in the forming-limit diagram under stretch-loading conditions.

The effect of ballistic-testing variables upon the residual velocity of projectiles and the absorbed energy in AA1100-H12 was investigated[65] by using design-of-experiment and artificial neural network approaches. Simulations were carried out using 3 process variables: the projectile nose-shape, the impact velocity and the target thickness. The Taguchi technique was used in designing experiments, with an L_{18} (61 x 33) orthogonal array of 18 combinations of test variables. The optimum test-variable combination was found by analysis of the signal to noise ratio. Simulated and experimental results agreed well, and the predictions of an artificial neural network model were in good agreement with experimental data.

Artificial intelligence methods were used[66] to determine the change in mechanical properties of AA2024 after aging for various temperatures and times. The hardness and bending strength were predicted on the basis of existing experimental data by using convolutional neural networks, artificial neural networks and random forest regression. It was found that the properties of aluminium alloys could be successfully determined by using artificial intelligence methods. The best results for powder-metallurgy AA2024 were an RMSE of 0.09068, an R^2 of 0.93476 and an MAE of 0.06734 by using the convolutional neural network algorithm. The best results for fully-dense AA2024 were an RMSE of 0.08578, an R^2 of 0.94166 and an MAE of 0.06212, found using the same algorithm.

Machine-learning methods, including neural networks, boosted trees, random forests, support-vector machines and k-nearest neighbours, were used[67] to plot diagrams for the fatigue fracture of D16T aluminium alloy, under regular loading. The boosted trees algorithm was based upon the recurrence partition of an input set into sub-sets which were associated with classes, and comprised the building and cancellation of trees. When creating trees, splitting and termination of learning criteria were used in which, during cancellation, some branches were removed. The random forest concept was based upon the construction of an ensemble which was made up of a number of decision-making trees which learned independently. The final result was then that chosen by a vote of all of the trees in the ensemble. The support-vector machines and k-nearest neighbour procedure were the simplest procedures. In the former method the data were represented by points in space. For learned data, split into two categories, the learning algorithm created a model which assigned new data to a certain category. The method of k-nearest neighbours introduced a new object to the class which predominated over k nearest-neighbours of the learning sample, with the distance between k nearest-neighbours being regarded as Euclidean. The neural network was, as usual, a sequence of neurons which were connected to one another, with the neuron being a computational unit which obtains information, performs mathematical operations on the information and forwards it to another neuron. Each input of a neuron receives a number of signals as the output of another neuron. Every input signal is multiplied by a weight which is analogous to a synaptic force and, as a result, the input data changes during its transmission from one neuron to another. The sum of all of the results is then found and the level of neuron activation is determined. The fatigue crack growth rate was predicted for load ratios, R, of 0, 0.2, 0.4 or 0.6. The sample consisted of 300 elements; 70% of which were chosen randomly for learning purposes. The remaining 30% were used for estimating the quality of the predictions. The results were in good agreement with the experimental data. The neural network method gave the highest accuracy, with an error of 2.5%. The errors which were associated with the methods of boosted trees, random forests, support-vector machines and k-nearest neighbours were 7.9, 12.9, 6.7 and 5%, respectively.

Table 11. Statistical analysis of fatigue crack growth predictions for AA7055-T7511

Technique	MAE	RMSE	R^2
DLFCO-DNN	7.11×10^{-5}	9.34×10^{-8}	0.1213
MFA-DNN	2.98×10^{-4}	7.97×10^{-8}	0.1976
GWO-DNN	3.09×10^{-5}	2.65×10^{-6}	0.2015
PSO-DNN	3.85×10^{-4}	1.01×10^{-7}	0.2768

DLFCO: chimp optimization algorithm with Levy flight random movement, GWO: grey wolf optimization algorithm, DNN: deep neural network, particle swarm optimization, MAE: mean absolute error, RMSE: root mean square error

Conventional methods for crack-growth prediction employ stress-analysis models and crack-growth model which are governed by the Paris law. When applied to long-term crack-growth prediction, they produce non-optimum solutions. Metaheuristic optimization algorithms were based upon neural networks in order to forecast more accurately the crack growth rates in aluminium alloys[68]. A dynamic Levy flight function was incorporated into a chimp optimization algorithm in order to train the deep neural network accurately. The chimp optimization algorithm is based upon simulated chimpanzee breeding and cluster-hunting. Their actions consist of attacking, chasing, barring and driving; all described by Volterra-Lotka-like equations. The 4 best results of optimum capability are chosen, and the remaining inhabitants upgrade their location with help from the data provided by the 4 best results. Levy flight is a random walk which is used for position-updating in a search for the global maximum position in the area of search. It has a non-Gaussian distribution which is based upon arbitrary numbers. The flight is chaotic and comprises both long and short jumps within the search space. The grey wolf optimization algorithm enhances accuracy. The performance of the predictive models was tested using AA7055-T7511 and AA6013-T651 data (figures 16 and 17), tables 11 and 12). This showed that the proposed predictive models offered a lower correlation error, least relative error, mean absolute error and root mean square error while shortening the run-time by 11.28%. The crack length and growth rates were predicted with high reliability and very high resolution.

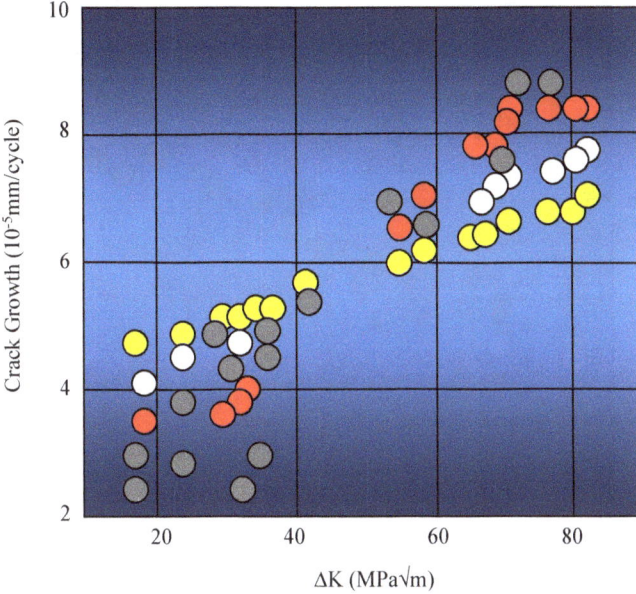

Figure 16. Predictions of the fatigue crack growth rate of AA7055-T7511. Red: chimp optimization algorithm with Levy flight random movement, white: grey wolf optimization algorithm, yellow: particle swarm optimization, grey: experiment

Table 12. Statistical analysis of fatigue crack growth predictions for AA6013-T651

Technique	MAE	RMSE	R²
DLFCO-DNN	7.14×10^{-5}	7.33×10^{-6}	3.48×10^{-4}
MFA-DNN	4.26×10^{-4}	1.21×10^{-4}	4.22×10^{-4}
GWO-DNN	5.65×10^{-4}	5.85×10^{-5}	5.04×10^{-4}
PSO-DNN	9.23×10^{-4}	1.99×10^{-4}	5.47×10^{-4}

DLFCO: chimp optimization algorithm with Levy flight random movement, GWO: grey wolf optimization algorithm, DNN: deep neural network, particle swarm optimization, MAE: mean absolute error, RMSE: root mean square error

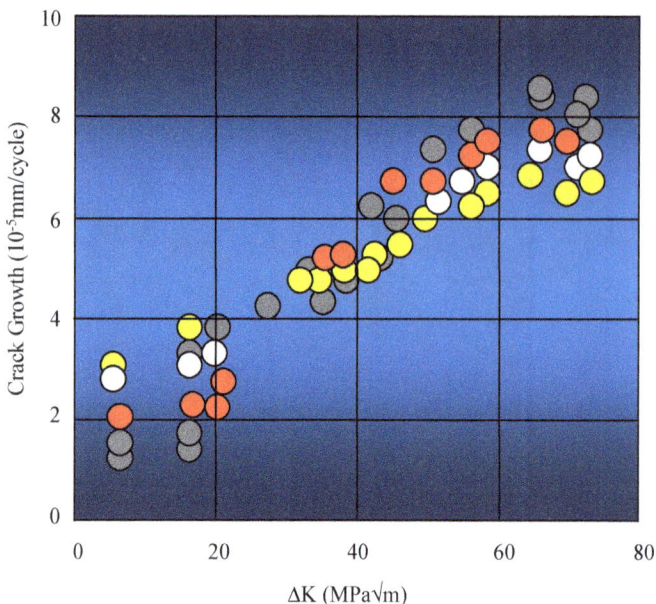

Figure 17. Predictions of the fatigue crack growth rate of AA6013-T651. red: chimp optimization algorithm with Levy flight random movement, white: grey wolf optimization algorithm, yellow: particle swarm optimization, grey: experiment

The fatigue life of AA2090-T83 alloy for constant amplitude and negative stress ratios was predicted[69] by using artificial intelligence and machine-learning techniques such as artificial neural networks, adaptive neuro-fuzzy inference systems, support-vector machines, the random forest model and an extreme-gradient tree-boosting model (figure 18). These were trained by using numerical and experimental input data which were obtained from fatigue tests involving a relatively small number of stress measurements. The coefficients of the traditional force-law formula were found by using numerical methods. As compared with traditional methods, the neural network and neuro-fuzzy models produced better results.

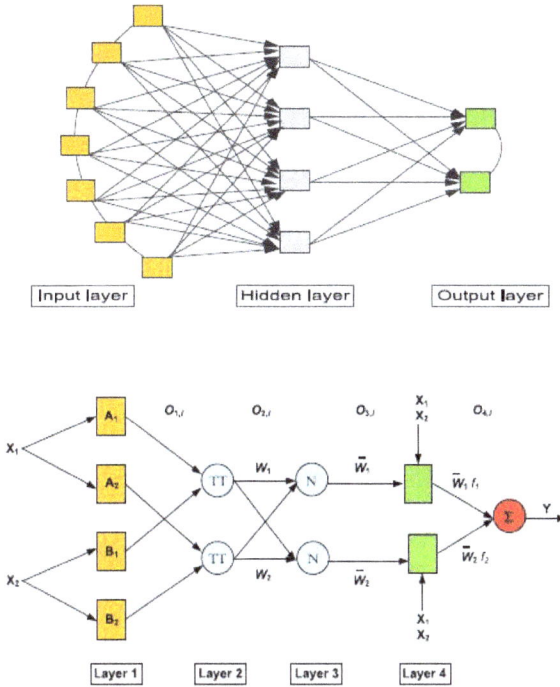

Figure 18. Artificial intelligence methods used. Top: multi-layer perceptron-based feed-forward, back-propagation network, bottom: ANFIS structure. Reproduced from: Fatigue life estimation of high strength 2090-T83 aluminium alloy under pure torsion loading using various machine learning techniques, M.S.Abdullatef, F.N.Alzubaidi, A.Al-Tamimi, Y.A.Mahmood, Fluid Dynamics and Materials Processing, 19[8] 2023, 2083-2083 under Creative Commons licence.

Table 13. Accuracy of results and duration of calculations for fatigue life predictions

Model	RMSE (experimental data)	RMSE (numerical data)	Duration (ms)
FF2	0.0025	0.0036	18.6
CF26	0.0017	0.0026	22.2
CF29	0.0026	0.0032	27.2
EL2	0.0025	0.0034	30.8
ANF1	0.00102	0.0061	164.0
ANF2	0.0078	0.0099	1793.1
SVR2	0.0026	0.0031	3.9
SVR3	0.023	0.027	2.3
SVR9	0.019	0.022	6.1
SVR17	0.021	0.024	7.9
RF	0.022	0.033	32.0
XGB	0.00165	0.0019	14.8

RMSE: root mean square error, CF: cascade-forward neural network, EL: Elman network, ANF: adaptive neuro-fuzzy system, SVR: support-vector machine model, RF: random forest model, FF: feed-forward neural network, XGB: extreme-gradient tree-boosting model

Neural network models which were trained by using the boosting iterations technique provided the best performance. By building strong models from weak models, the extreme-gradient tree-boosting model helped to predict the fatigue life by reducing model partiality and variations in supervised learning. Fuzzy neural models could be used to predict the fatigue life more accurately than could neural networks or traditional methods. The accuracy of the CF26 neural network findings, when compared to all of the models used in artificial neural networks, was notable. Four different kernel functions were considered for the support-vector machine models (table 13). Grid searches of hyperparameter values were performed for all four models. The Gaussian kernel function (SVR9) model out-performed the other models, in terms of accuracy, by an order of magnitude. The best model from each of the categories (CF26, ANF1, SVR9, XGB) was selected. The accuracy of each model was acceptable and consistent, but it was found that XGB offered better accuracy within an acceptable computational time than did the other models.

Table 14. Wear-rate of Al 4043 alloy reinforced with nanosize tungsten carbide, predicted using an artificial intelligence regression method (experimental values in bold)

COF	Rate(mm^3/Nm)	Friction Force(N)	COF	Rate(mm^3/Nm)	Friction Force(N)
0.21	**2.45**	**1.2**	0.23	2.29	1.14
0.21	**2.45**	**1.2**	0.23	2.2	1.14
0.21	**2.45**	**1.2**	0.23	2.29	1.14
0.26	**0.61**	**2.6**	0.27	1.25	3.16
0.26	**0.61**	**2.6**	0.27	1.25	3.16
0.26	**0.61**	**2.6**	0.27	1.25	3.16
0.28	**0.29**	**5.3**	0.30	0.13	5.23
0.28	**0.29**	**5.3**	0.30	0.13	5.23
0.28	**0.29**	**5.3**	0.30	0.13	5.2
0.22	**0.53**	**1.98**	0.15	0.46	1.34
0.21	**0.51**	**1.5**	0.15	0.46	1.34
0.21	**0.52**	**1.5**	0.15	0.46	1.34
0.06	**0.98**	**1.5**	0.16	0.87	2.34
0.26	**0.98**	**2.6**	0.16	0.87	2.34
0.26	**0.97**	**2.6**	0.16	0.87	2.34
0.01	**1.72**	**1.98**	0.21	1.2	3.5
0.63	**1.74**	**7.43**	0.21	1.20	3.52
0.15	**1.73**	**3.2**	0.21	1.20	3.52
0.01	**0.31**	**0.5**	0.04	0.15	0.46
0.03	**0.35**	**0.8**	0.04	0.15	0.46
0.01	**0.33**	**0.8**	0.04	0.15	0.46
0.06	**0.16**	**1.1**	0.09	0.41	1.70
0.06	**0.17**	**1.1**	0.09	0.41	1.70
0.05	**0.19**	**1.3**	0.09	0.41	1.70

0.07	0.52	2.1	0.10	0.82	2.70
0.11	0.53	2.3	0.10	0.82	2.70
0.13	0.55	3.32	0.10	0.82	2.70

Minimization of pressure die-casting defects in A380 alloy was investigated[70] by using artificial intelligence in the form of genetic algorithms and fuzzy logic. An L_{18} orthogonal-array Taguchi approach to the number of defects per batch in a lot of 500 components was chosen as a knowledge base. Analysis of variance was used to estimate the significance level in terms of the percentage contributions of input parameters and defects during pressure die-casting. The solidification time and the filling time were the most sensitive parameters. The melting-point, the plunger-velocity and the inoculation pressure made comparatively low percentage contributions. A genetic algorithm was used to optimize the process parameters, and the fitness value was calculated using fuzzy logic. By combining these artificial intelligence approaches, the optimum combination of process parameters was deduced.

The effect of cross-over fraction parameters upon the objective function value in the genetic algorithm optimization of machining parameters was assessed[71] with regard to the single-point diamond turning of RSA431 alloy. The cutting speed, feed-rate and depth of cut were optimized so as to give minimum surface roughness. A genetic model fitness function was developed using multiple regression modelling in order to establish the quality of each population member. Validation of the model was based on prediction accuracy. There was a linear relationship between the cross-over fraction parameters, giving a prediction accuracy of 95.05%.

The surface roughness in end-milling was modelled[72] using an adaptive neuro-fuzzy inference system and genetic algorithms. The spindle speed, the feed-rate, the depth-of-cut and the workpiece-tool vibration amplitude were used as inputs to model the workpiece surface roughness. The parameters used in the model, together with the most relevant inputs, were selected by using genetic algorithms which maximised modelling accuracy. The trained model was tested by using the set of validation data and the procedure was illustrated by using experimental data from CNC end-milling of AA6061. The results demonstrated the effectiveness of the approach in modelling surface roughness.

Table 15. Wear of Al 4043 alloy reinforced with nanosize tungsten carbide, predicted using an artificial intelligence neural network method (experimental values in bold type)

COF	Rate(mm³/Nm)	Force(N)	COF	Rate(mm³/Nm)	Force(N)
0.21	**2.45**	**1.2**	0.20	2.4477	1.1985
0.21	**2.45**	**1.2**	0.20936	2.4477	1.1985
0.21	**2.45**	**1.2**	0.20936	2.4477	1.1985
0.26	**0.61**	**2.6**	0.24738	0.59965	2.6066
0.26	**0.61**	**2.6**	0.24738	0.59965	2.6066
0.26	**0.61**	**2.6**	0.24738	0.59965	2.6066
0.28	**0.29**	**5.3**	0.31241	0.32606	5.2961
0.28	**0.29**	**5.3**	0.31241	0.32606	5.2961
0.28	**0.29**	**5.3**	0.31241	0.32606	5.2961
0.22	**0.53**	**1.98**	0.20074	0.93881	1.6674
0.21	**0.51**	**1.5**	0.20074	0.93881	1.6674
0.21	**0.52**	**1.5**	0.20074	0.93881	1.6674
0.06	**0.98**	**1.5**	0.18243	0.99874	2.093
0.26	**0.98**	**2.6**	0.18243	0.99874	2.093
0.26	**0.97**	**2.6**	0.18243	0.99874	2.093
0.01	**1.72**	**1.98**	0.3399	1.5002	5.2805
0.63	**1.74**	**7.43**	0.3399	1.5002	5.2805
0.15	**1.73**	**3.2**	0.3399	1.5002	5.2805
0.01	**0.31**	**0.5**	0.024652	0.32558	0.66947
0.03	**0.35**	**0.8**	0.024652	0.32558	0.66947
0.01	**0.33**	**0.8**	0.024652	0.32558	0.66947
0.06	**0.16**	**1.1**	0.07686	0.23104	1.1952
0.06	**0.17**	**1.1**	0.07686	0.23104	1.1952
0.05	**0.19**	**1.3**	0.07686	0.23104	1.1952

0.07	0.52	2.1	0.1194	0.54476	2.2662
0.11	0.53	2.3	0.1194	0.54476	2.2662
0.13	0.55	3.32	0.1194	0.54476	2.2662

The Al 4043 alloy was reinforced with various (1, 3, 5%) fractions of nanosized tungsten carbide in order to increase wear resistance[73]. A Taguchi L27 orthogonal array was employed for wear analysis, and the Taguchi signal-to-noise ratio was used to deduce the optimum parameters which minimized the wear and the coefficient of friction (COF). A regression model and an artificial neural network were used to predict the experimental results (tables 14 and 15). It was deduced that the optimum combination was a specific wear-rate of 0.12mm³/Nm, a coefficient of friction of 0.01 and a frictional force of 1.02N.

Drilling conditions such as the feed-rate, the cutting-speed and the drill diameter, were introduced into an artificial neural network in order to predict the formation of burrs during the drilling of AA7075[74]. The main exit-burr characteristics, of size and type, were classified experimentally. Three types of exit-burr; uniform, uniform with cap and transient burrs, were observed. Acoustic emission sensor monitoring was used to sample time-series signals during burr formation. The burr types could later be predicted by machine-learning and deep learning techniques. Back-propagation neural networks were constructed by using the drilling conditions and the acoustic emission signals as input vectors. A convolution neural network was used to obtain spectrogram-image inputs from the acoustic data. The convolution neural network predicted burr types with better accuracy than did the back-propagation network model; e.g., 0.9375 rather than 0.8571.

A study was made[75] of the relationship between the water contact-angle and the composition, surface finish, and microstructure for aluminium alloys with a silicon content of 7 to 50%, and aluminium-matrix composites with graphite, $NiAl_3$, and SiC. The surface properties were modified by mechanical abrasion, etching and the addition of alloying elements. Machine-learning was used to investigate correlations between the predictor variables (material properties) and the contact angle. Theoretical models for the wetting of rough surfaces did not fully explain the contact angle, while machine-learning models followed the experimental values. A full factorial study was used, together with combinations of all levels of the predictor factors: grit size, silicon content, droplet size, elapsed time, etching, reinforcing particle. In order to map the predictor variables to the response variables, 409 experimental data points were used to train and test supervised machine-learning models: regression, artificial neural network, chi-squared automatic interaction detection, extreme gradient boost, random forest. The most accurately trained model offered a strong positive linear correlation (R > 0.9) between predicted and observed contact angles. The experimental measurements and artificial intelligence results showed that the contact angle was increased by abrading the surface, etching or adding graphite.

Copper Alloys

The equation for the flow-curve of Cu-Cr-Zr alloys was studied[76] by using genetic programming; the evolutionary optimization technique based upon the Darwinian evolution of species. The principal characteristic of genetic programming is its non-deterministic behaviour. A comparison of experimental results, analytical solutions and the genetic algorithm showed that the genetic programming method was very promising.

A model based upon least-squares support vector machines was capable of forecasting the hardness of Cu-3Ti-1Cr alloy[77]. In order to improve the predictive accuracy, the leave-one-out cross-validation technique was chosen to determine automatically the optimum hyper-parameters of the least-squares support vector machine method. The forecasting performance of the latter model and the partial least squares regression method integrated with the radial basis function technique was compared with experimental values. This showed that the least-square support vector machine model was better than the conventional radial basis function plus partial least squares model for predicting the hardness of the Cu-3Ti-1Cr alloy. The calculated results were consistent with experimental values. The least-squares support vector machine method was expected to be able to predict the variation of the hardness of copper alloys as a function of prior cold work, aging temperature and aging time.

Genetic programming was used to predict the grain-size of electrodeposited $Cu_{1-x}Zn_x$ alloys as a function of the zinc and copper contents of the electrolyte and alloy films[78]. In order to predict the grain-size, 48 different competitors were considered. Each of them differed with regard to linking function, number of genes, head size and chromosomes. In order to generate databases for the new grain-size competitors, testing and training sets among a total of 134 samples were selected for various zinc and copper ratios. The testing and training sets consisted of randomly selected 106 and 28 cases. Six input parameters were chosen, such as the zinc and copper contents in the electrolyte and thin films. The output was the grain-size of the electrodeposited alloy. The model showed that all of the input parameters affected the grain-size.

Porous alloy-composites are useful for the grinding of superalloys. The flexural strength is an important mechanical property which is associated with the porosity level and with inhomogeneity in such composites. Due to the non-linearity of the constituents of the composite, the prediction of mechanical properties by using conventional regression models can be unsatisfactory. An evaluation was made[79] of the efficacy of artificial neural networks in predicting the flexural strength of porous Cu-Sn-Ti composites containing molybdenum disulfide particles. The input parameters for the artificial neural network were the average particles size, the porosity volume and the weight fraction of MoS_2 particles. The determination of the number of hidden neurons in the single hidden layer of the artificial neural network model was obtained via empirical formulation. The neural network model was compared with a conventional multiple linear regression model. The neural network could better predict the flexural strength of the porous composite.

Machine-learning algorithms, trained using experimental data, were used to predict the sintered density of powder-metallurgy samples[80]. The inputs included the processing parameters, the alloy composition and the properties of the raw materials. A multilayer perceptron model out-performed 4 other regression and neutral network models, with a high coefficient-of-correlation and low error. The sintered density which was predicted by the machine-learning model agreed well with experimental data; with an error of less than 0.028. The machine-learning model was applied to Cu-9Al alloy and was used to select the processing parameters which were required in order to attain the expected sintered density of 0.88. The powders were treated by using the predicted particle size, pressing pressure and sintering temperature, giving a relative sintered density of 0.885.

Artificial intelligence methods were used[81] to determine the relationship between the thickness and the forming rate of cups drawn from electrolytic tough pitch copper sheet. Machine-learning techniques were expected to be able to predict the forming limit diagram of copper alloys. The aim here was to create machine-learning artificial neural network tools for the modelling of the relationship between the thickness and forming-rates as a function of formability. The forming limit diagram was determined for copper strips having initial dimensions of 1500mm in length, 750mm in width and 6mm in thickness. The thickness was reduced by 50% in 9 successive incremental steps. Sheets of 3, 1.5, 0.75, 0.38 and 0.19mm were thus obtained and used to determine the forming limit diagram via the Nakajima approach. A finite-element model of the drawing was created and the simulation results were used to train a 2-step machine-learning model. A Bayesian regularization and Levenberg-Marquardt algorithm were first used to predict the maximum and minimum points of stress. The minor strains which were predicted in the first step were then used as inputs. Using the same feature-set, the Bayesian regularization and Levenberg-Marquardt algorithm predicted the major strain. This exhibited a linear trend up to the middle and then became non-linear. The trained machine-learning model was used to predict unknown intermediate values for 2mm- and 0.25mm-thick sheets. The difference between the forming limit diagrams according to predicted and experimentally verified data fell between 2 and 5%.

High-Entropy Alloys

This is a field in which artificial intelligence can really be expected to shine, given that the compositions severely flout the venerable rules of Hume-Rothery, thus leaving traditional metallurgists with little alternative guidance. The high-entropy alloys constitute a paradigm-change in materials science but one which is slowed by traditional thinking. Vapour–crystal transformation, for example, creates nanoparticles of alloys existing in sometimes unprecedented states. Any resultant high-entropy alloy nanostructures may be impossible to produce by means of other additive manufacturing techniques, because of miscibility limits. Their exceptional properties were originally thought to be related to entropy maximization but entropy, as the only driving force, may be insufficient to permit alloying via traditional liquid-solid transformation.

There is now a sufficient volume of experimental data on multi principal element alloys that one can hope to discover connections between the elemental properties and phases such as monophase solid solutions and amorphous or intermetallic compounds. In order to obtain such insights, neural networks in the machine-learning framework have been used[82] to detect underlying data-patterns in which an experimental data-set is used to classify the corresponding phase selection in the alloys. By using the full data-set, a trained neural network model could attain an accuracy of over 99%; that is, more than 99% of the phases in the alloys were correctly identified. The trained neural network parameters suggested that the valence electron concentration played the predominant role in determining the phases. Cross-validation training and test data-sets revealed an average accuracy of better than 80%.

Bayesian optimization is already a powerful method for exploring and exploiting material-design spaces. A variant was proposed[83] which was capable of actively learning material-design as a multiple objective and constraint problem. It was demonstrated by exploring the space of refractory multi-principal-element alloys; in particular, the Mo-Nb-Ti-V-W system. The alloys were explored in order to optimize two density-functional theory derived ductility indicators: the Pugh ratio and the Cauchy pressure. Alloys at the Bayesian optimization Pareto-front were analyzed using density functional theory in order to derive a fundamental basis for their superior performance.

An explainable artificial intelligence approach was used[84] to predict rapidly, and explain, trends in the temperature-dependent yield strength of as-cast multi-principal element alloys. This was based upon an ensemble of support vector regression models. The interpretation of trained support vector results was based upon Shapley additive explanation. This revealed trends that explained the variation in yield stress as a function of temperature, component elements and phases present. Alloys having the highest room-temperature yield stresses were expected to occur in multi-phase refractory-metal based 6-component systems, in which the microstructure comprised secondary-phase precipitates within a body-centred cubic matrix. The probability of secondary-phase formation was increased by optimum concentrations of aluminium and/or silicon additions. Even when the alloys had a face-centred cubic structure, secondary phases were still the key to ensuring a high yield stress at room temperature. Two additional factors which favoured a high room-temperature yield stress was a composition having equiatomic or near-equiatomic concentrations and alloying elements possessing an optimum number of p-electrons in the valence configuration. High-temperature yield-stress trends which were associated with single-phase and multi-phase body-centred cubic alloys were dominated by the temperature factor, and this obscured the effect of other variables.

The adaptive neuro-fuzzy interface system was used[85], for the first time, to predict the phases of high-entropy alloys via two differing approaches. One involved using the component elements as input parameters. The other involved 6 parameters which are pivotal in the formation of the alloys. The accuracies of the models were confirmed by using a test data-set and were found to be 84.21% and 80%, respectively.

It was shown[86] that random forest models, when trained using unconstrained maximum depths, can often report some randomly generated factor as being one of the most critical factors in generating predictions which classify an alloy as being of high-entropy type. This is the case for impurity, permutation and Shapley importance rankings. It was demonstrated that, in the case of impurity importance rankings, optimizing only the validation accuracy yields models which prefer the random feature when generating predictions. It was noted that, by choosing a Pareto optimization strategy to balance validation statistics with the differences between the training and validation statistics, models that reject random features are obtained.

A transferable machine-learning model was proposed[87] which was based upon the intrinsic properties of substrates and adsorbates. It could predict the adsorption energies of single-atom alloys, AB intermetallics and high-entropy alloys simply by training for the properties of transition metals. This model deduced the structure-activity relationship of the adsorption energies of alloys from the viewpoint of machine-learning. This revealed the role played by surface-atom valence, electronegativity and coordination, and by the adsorbate valence, in determining adsorption energies.

Genetic algorithms were used as a factor-selection method for predicting the hardness of high-entropy alloys[88]. The concepts of factor-importance and gene manipulation were incorporated into a genetic algorithm so as to make it more comprehensible. Comparison showed that the improved genetic algorithm was better than the traditional genetic algorithm with regard to accuracy, stability and efficiency. Machine-learning model selection was considered, with composition factors being selected by the modified genetic algorithm. In order to improve the predictive ability of machine-learning, the stacking method as an ensemble learning strategy was proposed for Al-Co-Cr-Cu-Fe-Ni high-entropy alloy hardness prediction. The errors were indeed lowered.

Table 16. Machine-learning hardness predictions for high-entropy alloys

Composition	Experimental (HV)	Predicted (HV)	Error (%)
$Fe_{22.5}Co_{25}Ni_{25}Cr_{20}V_5Nb_{2.5}$	215	212.15	1.32
$Fe_{20}Co_{25}Ni_{25}Cr_{20}V_5Nb_5$	275	277	0.72
$Fe_{17.5}Co_{25}Ni_{25}Cr_{20}V_5Nb_{7.5}$	501	375.96	24.95
$Fe_{15}Co_{25}Ni_{25}Cr_{20}V_5Nb_{10}$	510	487.11	4.48
$Fe_{32.5}Co_{10}Ni_{25}Cr_{15}Mn_5V_{10}Nb_{2.5}$	277	268.47	3.07
$Fe_{30}Co_{10}Ni_{25}Cr_{15}Mn_5V_{10}Nb_5$	315.16	332.37	5.46
$Fe_{27.5}Co_{10}Ni_{25}Cr_{15}Mn_5V_{10}Nb_{7.5}$	449.16	423.9	5.62
$Fe_{25}Co_{10}Ni_{25}Cr_{15}Mn_5V_{10}Nb_{10}$	607.6	533.26	12.23
$Fe_{32.5-x}Co_{10}Ni_{25}Cr_{15}Mn_5V_{10}Al_{2.5}Nb_5$	318.42	464.12	45.75
$Fe_{32.5-x}Co_{10}Ni_{25}Cr_{15}Mn_5V_{10}Al_{2.5}Nb_{7.5}$	519	573.48	10.49
$Fe_{32.5-x}Co_{10}Ni_{25}Cr_{15}Mn_5V_{10}Al_{2.5}Nb_{10}$	589	682.85	15.93
$Fe_{32.5-x}Co_{10}Ni_{25}Cr_{15}Mn_5V_{10}Al_{2.5}Nb_{12.5}$	613.9	792.21	29.04

Predicting the phases of high-entropy alloys is clearly complex as it depends upon a large number of thermal, physical and geometrical parameters. The phases can include intermetallic and amorphous forms. Because the trial-and-error approach is time-consuming, both experimentally and numerically, XGBoost, random forest, AdaBoost, decision tree, logistic regression, support vector machine and k-nearest neighbour methods have been adopted[89]. The XGboost algorithm yields the better accuracy (90 %). Machine learning methods well predicted the phases and also identified the main significant variables such as mean atomic radius, atomic size difference, mixing enthalpy, ideal mixing entropy and valence electron concentration, which are in consistence with experimental results. As a whole, it is recommended that the widespread use of machine learning approaches could accelerate HEA process with tailored configurations. The use of multiple models in ensemble learning reduces the risk of over-fitting, increases the accuracy of predictions, and enhances the generalizability of models.

Five machine-learning models were applied[90] to the prediction of the phases present in $Fe_{25-x}Co_{25}Ni_{25}Cr_{20}V_5Nb_x$ high-entropy alloys with x = 2.5, 5, 7.5 or 10at %. The models were: decision tree, random forest, bagging, AdaBoost and extra trees. In addition, 6 machine-learning models, XGBoost, gradient boost, bagging, extra trees, random forest and artificial neural network, were used to predict the hardness. To recapitulate: bagging and boosting are ensemble learning techniques which improve the accuracy of results by combining various algorithms. Bagging reduces the variance. Boosting is intended to reduce bias by combining weak classifiers. It is a successive training process in which learners learn from previous learner and reduce error by adjusting the weights. Extra trees is another ensemble learning method which builds multiple decision trees over the complete data-sets and randomizes the split. It is faster, involves low computational cost and requires less time, due to the random selection of splits. Factors were identified which determine the phases which appear in the alloy (figure 19), as were those which governed the hardness (figure 20). The order of importance of the factors which govern the appearance of face-centred cubic plus intermetallic phases is seen to be: Fe > melting-point > VEC > Ni > Nb > Zr and Ti > Hf > Ta > Co and Mn > $M_{enthalpy}$ > Mo > V > Cr > Al > ρ > Cu > $M_{entropy}$. Melting-point, with an importance of -0.37, has a large effect upon FCC+IM formation. The negative sign indicates that decreasing its value would favour FCC+IM phase formation. The melting-point of an alloying element is characteristic of its bonding energy. A higher bond-energy suggests a higher melting-point for the system. A constituent element with a higher melting-point would not typically allow the inclusion of a metallic atom with a lower bonding energy, leading to multiple phases. With regard to hardness, it can be seen that the order of importance of the factors which govern hardness Fe > Ti (Mo, Mn) > Nb > Ni > V > Zr > Al > Co > Cu. The maximum correlation between hardness and iron content is about -0.37. The negative sign shows that lowering the iron content could increase the hardness. Titanium and molybdenum have a positive effect upon hardness, so that increasing their contents would also increase hardness. Following evaluation of the performance of each model, an extra trees classifier which offered an accuracy of 0.893 and an artificial neural network which offered an R^2 of 0.95 and an MAE of 34.91, were deemed to provide the best predictive capabilities for phase type and hardness, respectively. The machine-learning models were used to predict the phase type of 16 high-entropy alloys and the hardness of 12 alloys before checking the results by experiment. The extra trees model successfully identified the phases of the present alloy and also of other high-entropy alloys (table 16). The artificial neural network model predicted the experimentally measured hardness with an average error of 13.25% (figure 21).

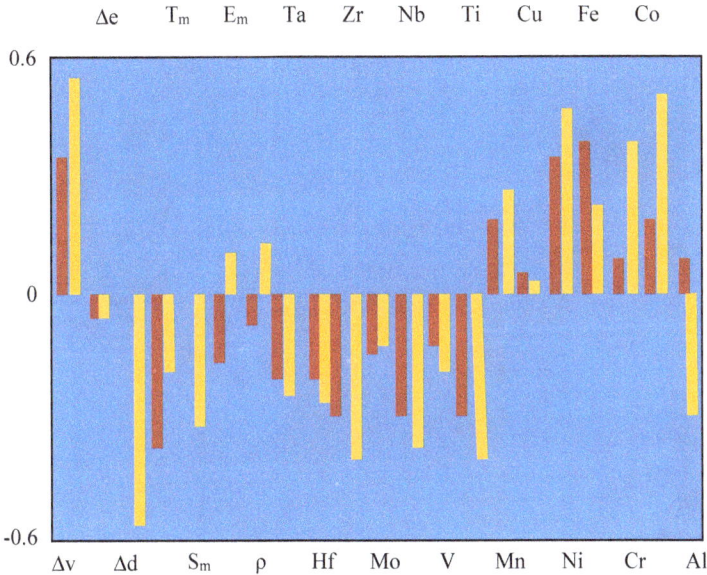

Figure 19. Factors governing the phases in high-entropy alloys. Δv: valence-electron difference, Δe: electronegativity difference, Δd: atomic size-difference, T_m: melting point, S_m: mixing entropy, E_m: mixing enthalpy, ρ: density. red: face-centred cubic, yellow: face-centred cubic plus intermetallics

Table 17. Comparison of the predicted and experimental HEA hardness values

Composition	Predicted Value (HV)	Experimental Value (HV)
$Co_{10}Cr_{20}Fe_{30}Ni_{40}$	121.56	126.5
$Co_{10}Cr_{20}Fe_{40}Ni_{30}$	123.4	123.4
$Al_{41}Co_{20}Cr_{19}Fe_{15}Ni_5$	791	762.8
$Al_{46}Co_{16}Cr_{15}Fe_{15}Ni_8$	873	701.25
$Al_{32}Co_{13}Cr_{33}Fe_{22}$	727.2	735.5

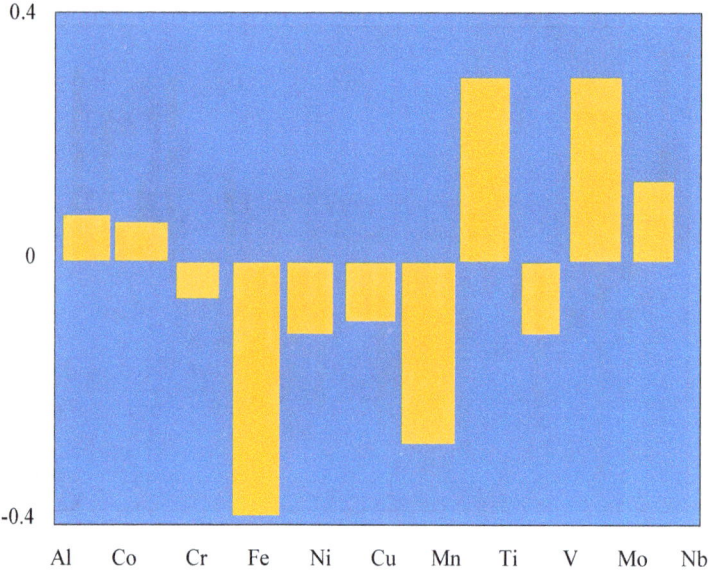

Figure 20. Factors governing the hardness of high-entropy alloys.

A machine-learning model was based[91] upon hardness data for high-entropy alloys in the Al-Co-Cr-Cu-Fe-Ni system. This involved the use of random forest, XGBoost, lightGBM and CatBoost and the stacking ensemble algorithm. The R^2 value of the model attained 0.93 following ten-fold cross-validation. The ensemble learning approach was stable and accurately predicted the hardness, as experimentally verified (table 17). The model could be applied to other high-entropy alloy systems, as well as to low-hardness Cr-Fe-Ni medium-entropy alloys. It was noted that additional aluminium and chromium could increase the hardness, while adding high atom-ratios of nickel, cobalt and iron could suppress the hardness. Valence-electron concentrations below 6.7 could promote the formation of body-centred cubic phases and improve hardness. The δG value had a positive effect upon hardness, while a low H_{mix} or a high S_{mix} value was beneficial in increasing the hardness.

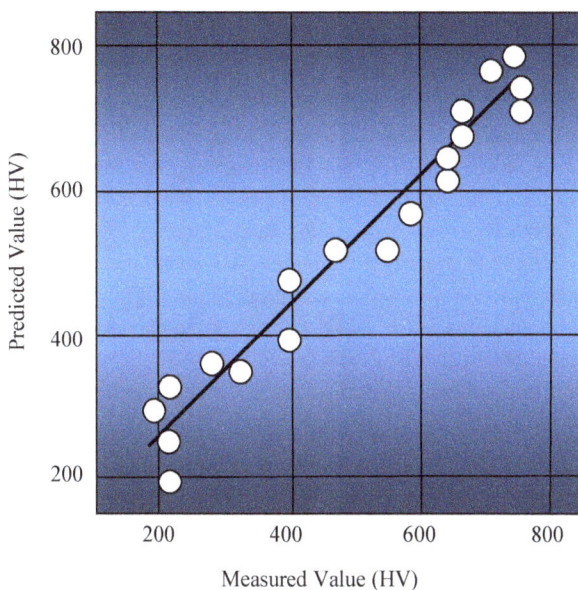

Figure 21. Prediction of the hardness of high-entropy alloy using an artificial neural network

Iron Alloys

A back-propagation neural network model for predicting the mechanical properties of low-alloy nickel-chromium steels, and the effect of each addition upon the properties, was developed[92]. Simulated results showed that the average output-prediction error exhibited by the back-propagation network was lower than 5% of the prediction range in more than 95% of the cases.

Artificial neural networks were used[93] to model cold alloy-steel bars, and correlate heat-treatment parameters with end-product quality. The multi-layered feed-forward neural networks enabled the mapping of inputs such as physical dimensions, material composition and heat-treatment parameters to the Brinell hardness and the UTS. These properties were checked by using new data-sets and this revealed a satisfactory level of agreement for a range of conditions. The neural networks were integrated into a genetic algorithm search strategy in order to identify the best material characteristics and furnace-operating conditions for maximizing the hardness and strength[94].

The compressive deformation of 42CrMo steel was investigated[95] at 850 to 1150C and strain-rates of 0.01 to 50/s using a Gleeble-1500 machine. An artificial neural network was developed in order to predict the constitutive flow behaviour of the steel during hot deformation. The inputs were the deformation temperature, the logarithmic strain-rate and the strain, while the flow stress was the output. A 3-layer feed-forward network with 12 neurons in a single hidden layer, and a back-propagation learning algorithm was used. The effect of deformation temperature, strain-rate and strain upon the flow behaviour was compared with experimental data and the predicted results were consistent with what was expected on the basis of the fundamental theory of hot compressive deformation.

The effect of the cold-working temperature, the degree of deformation, the strain-rate and the initial austenite grain size upon the volume fraction of strain-induced martensite in AISI301 stainless steel was modelled[96] by using artificial neural networks. The results of the network model agreed well with published experimental data. The best ranges of the processing parameters for grain-refinement via strain-induced martensitic transformation and reversion to austenite were deduced by using the model.

Machine-learning models have been recommended[97] for determining glass-forming ability without tedious experimentation. Glass-forming ability can be predicted by using random regression forest, extreme gradient boosting algorithm, extra regression tree, k nearest-neighbour, support vector machine, gradient boosting decision tree, decision regression tree and artificial neural network methods. Studies confirmed that the extreme gradient boosting model is more efficient than the artificial neural network model, due to its high accuracy and precision. A machine-learning method was used[98] to predict glass formation in various alloys. The model accurately predicted glass-formation and the critical thickness of metallic glasses. The ternary Fe–B–Co system in particular was studied and the effects of minor additions of chromium, niobium and yttrium were evaluated. Minor additions of niobium and yttrium led to significant improvements in the glass-forming ability of the Fe–B–Co system, but there was a shift in the optimized alloy composition.

Machine-learning was used to optimize the amount of metalloid elements (Si, B, P) to be added to an iron-based amorphous alloy in order to improve the soft magnetic properties[99]. The effect of the metalloid elements was investigated by correlation analysis. Silicon and phosphorus affected the saturation magnetization while boron affected coercivity. The coefficient-of-determination, R^2, for regression analysis after learning using the random forest algorithm was 0.95 The R^2 value, after including phase information on Fe-Si-B-P ribbon, increased to 0.98.

Vacancy-migration energies, as functions of the local atomic configuration in Fe-Cu alloys were tabulated[100] by using a suitable interatomic potential for the given alloy. Sub-sets of the energies were then used to train an artificial neural network so as to predict all of the vacancy migration energies. The error in prediction was evaluated by using a fuzzy-logic system which allowed feedback to be exploited for further training and improved prediction. The system was used describe the kinetic path which was followed during the diffusion of solute atoms in the alloy via a vacancy mechanism. Iron-copper alloys were chosen because of the importance of copper precipitation in iron with regard to embrittlement. An atomistic kinetic Monte Carlo model, which took account of local chemistry and relaxation effects for assessing energy-barriers to point-defect migration, was applied[101] to the study of microchemical evolution as driven by vacancy diffusion in Fe-Cu and Fe-Cu-Ni alloys[102]. Copper-precipitation, enhanced by the presence of nickel is a major cause of hardening and embrittlement in such alloys. Local chemistry and relaxation effects were introduced by using artificial intelligence neural networks to calculate the barriers to vacancy migration as a function of the local atomic environment. The use of neural networks was equivalent to calculating the energy barriers by using nudged elastic bands methods with an interatomic potential, but the former method was more efficient.

Cyclic oxidation depends upon factors such as time, temperature, atmosphere (composition, pressure) and alloy composition. The CatBoost machine-learning algorithm was used[103] to model the cyclic oxidation of binary Fe-(10–20)% Cr alloys and ternary Fe-(16–20)Cr-(10–30)%Ni alloys by using published data. The oxidation conditions were 650 or 800C in air plus 10% of water vapour. The CatBoost model was successfully trained and tested by using a 80:20 ratio of the data-set, with the composition, the temperature and the cycle-time as the inputs and the mass change as the output. Five-fold cross-validation showed that the model had an average R^2 of 0.98. The CatBoost algorithm, when given a composition, could predict whether the alloy would form a protective oxide scale, would oxidize rapidly or exhibit spallation following 100h of cyclic oxidation.

A problem with the use of Fe-Cr-Al alloys at high temperatures is the formation of α'-precipitates, because these cause brittleness and fracture failure. Precipitation causes a hardness change during aging which depends upon the alloy composition, the treatment time and the temperature. A Gaussian process regression model was instructed using hardness data. A Shapley additive explanations algorithm was used as an artificial intelligence tool to comprehend the effect of various parameters in causing the hardness change. This confirmed[104] that the primary promoter of α' age-hardening in the Fe-Cr-Al system was chromium. The analysis also indicated that aluminium did not have a clear role in only suppressing the formation of α'; contrary to received opinion. The lack of an effect of aluminium upon age-hardening was attributed to the ability of aluminium both to suppress thermodynamically and enhance kinetically the formation of α'. The analysis also indicated that molybdenum had no clear tendency to enhance or suppress α'.

Various artificial intelligence tools were used[105] to model the pitting corrosion behaviour of EN1.4404 austenitic stainless steel when subjected to polarization tests in chloride solutions with precursor salts such as NaCl and $MgCl_2$. In order to determine the pitting potentials, various environmental conditions were imposed by varying the chloride-ion concentration, the pH value and the temperature. Several techniques were used: classification trees, discriminant analysis, k-nearest-neighbours, back-propagation neural networks and support vector machines. The results revealed a good correlation between experimental and predicted data for all of the cases studied[106].

Artificial intelligence was used[107] to design a series of creep-resistant steels by combining an integrated thermodynamics and kinetics based model with a genetic algorithm optimization technique. Steel compositions which contained some 9 elements were combined with suitable heat-treatment parameters so as to produce target microstructures (ferritic, martensitic, austenitic) for high-temperature applications. Solid-solution strengthening and precipitation-hardening were employed to improve the strengths of the designed steels. Their combinations were optimised for the 3 families of steel considered. Each model was checked by analysing the strengthening found in existing steels. Very good agreement was found between experimental data and the model predictions, and all of the new alloys were predicted to out-perform existing high-end steels.

A neural net model was based upon tests, performed on samples of experimental alloys, with regard to the dependence of the yield stress upon the chemical composition, the heat-treatment and the test temperature[108]. Calculations were repeated 150 times and frequency histograms of the distributions of the concentrations of elements were constructed. It was noted that the presence dissolved vanadium did not strengthen solid solutions, but vanadium could form highly-dispersed carbides, nitrides, and carbonitrides.

Oxide dispersion-strengthened steels were developed for which the mechanical properties could be improved by optimizing the chemical composition and heat-treatment[109]. The modelling of such alloys by analytical methods was insufficiently accurate and artificial intelligence methods such as machine-learning were needed. Three hybrid machine-learning techniques were used to estimate the UTS and elongation. These were a feed-forward artificial neural network trained using particle swarm optimization and 2 adaptive neuro-fuzzy inference system methods which were trained using both fuzzy C-means clustering and subtractive clustering. Because the oxide-particle hardened alloys were usually produced by mechanical alloying of a mixture of powders, followed by consolidation and hot-rolling, tensile tests were performed on divers variants of the alloys. The main result was to be able to estimate the UTS and elongation on the basis of factors such as the contents of aluminium, molybdenum, iron, chromium, tantalum, yttrium and oxygen and the mechanical alloying and heat-treatment conditions. The artificial intelligence methods could accurately determine the behaviour of the alloys, with an accuracy of about 95%.

Computational intelligence and knowledge-based techniques were used[110] to predict the flow stress of the high-strength low-alloy steel, AISI4340. The estimation process was based upon various choices for the steel's microstructure, as quantified by the hardness, and took account of the applied temperature, the strain and the strain-rate. The simulated results produced using both techniques showed good agreement with experimental data. Machining data could be used[111] to predict the cutting-force and optimize the processing parameters. Cutting force is an essential parameter that has a significant impact on the metal turning process. A cutting-force prediction model for the turning of AISI4340 alloy steel was developed by using Gaussian process regression, support vector machines and artificial neural networks. The Gaussian method provided reliable predictions of the surface roughness for dry-turning, with $R^2 = 0.9843$, a mean absolute percentage error of 5.12% and a root mean square error of 1.86%.

The machinability of the super-duplex stainless steel, EN1.4410, was investigated[112] with regard to its machinability in the absence of a cooling or lubricating medium. Experimental data were generated for a range of input parameters and analysed by choosing surface roughnesses as output data. Predictive and mathematical models were used to minimize the surface roughness. The influence of the input parameters upon the surface roughness was studied and the optimum values of the input parameters were deduced by using a genetic algorithm. The accuracy of the predictive models was checked against various sets of experimental data. Samples of a super-duplex stainless steel with a ferrite matrix and dispersed austenite particles were deformed[113] in torsion at 900 to 1200°C at strain-rates from 0.01/s to 10/s. The level and shape of the plastic flow-stress curves depended upon the temperature and the strain-rate and varied with the austenite volume fraction. When the two phases deformed together, the marked difference in the softening behaviours of the austenite and ferrite led to uneven strain-partitioning between the phases. The plastic behaviour of the two-phase material was therefore more complex than that of a single-phase material. Experimental data from hot-deformation testing provided inputs in the form of temperature, strain-rate, strain and the stress resulting from the material during deformation. These inputs were fed to machine-learning algorithms such as an artificial neural network with one hidden layer, or a neural network with a specialist system (ANFIS). Following machine learning, the plastic flow curves were rebuilt and compared with those obtained experimentally. The ability of the algorithms to rebuild the plastic flow curves of the super-duplex stainless steel was associated with changes in the shapes of the flow curves and microstructural evolution.

Deep machine-learning was used to predict the microsegregation behaviour in iron-based binary alloys with additions of carbon, silicon, manganese and phosphorus[114]. Training data for the machine-learning model were taken from quantitative phase-field simulations of directional solidification, thus taking account of the effects of microstructural evolution upon the microsegregation behaviour. The method could also be coupled with a macrosegregation model. The simulated results of the macrosegregation model were different to those obtained using a conventional macrosegregation Scheil model or a model constructed with training data from a 1-dimensional finite-difference calculation of the microsegregation. This emphasized the importance of accurately describing the microsegregation behaviour when predicting macrosegregation. It was concluded[115] that deep learning was a promising ad easy means for predicting microsegregation with high accuracy.

A data-driven machine-learning technique was proposed for mining the literature on experimental data concerning stacking-fault energies in steels[116]. The algorithms of machine-learning could be used to visualize stacking-fault energy data and then construct a 3-class classifier. The latter was then used to predict the probable secondary deformation mechanisms of untested compositions. The classifier itself was a valuable tool for developing austenitic steels in which specific secondary plastic deformation mechanisms are an important feature. The stacking-fault energy tends to be very dependent upon the chemical composition and temperature. Machine-learning was expected to be able to clarify this dependence.

A new approach was proposed[117] for the real-time adjustment of feed-rate and spindle-speed so as to improve the surface quality of machined parts. The controller posited was a model-based closed-loop system which comprised a surface-roughness prediction model and a multi-variable tool to ensure real-time improvements in surface quality during machining. A test study was based upon the milling of EN24T steel. The simulation results showed that the controller markedly reduced the disparity, between the required and predicted surface roughness, from 3.6μm to 0.12μm.

Data analysis, based upon decision tree and random forest machine-learning algorithms, was used[118] to predict when interphase precipitation occurs. Nanosized interphase precipitates, which form in ordered rows, are important if high-strength low-alloy steels, are to attain desired strengths. The occurrence of interphase precipitation, rather than random heterogeneous precipitation within grains, depends upon interconnected factors such as the alloy composition, ferrite growth-rate, temperature, crystallography and precipitate shape and size. The decision tree model was unable to provide an accurate prediction, while the random forest could achieve a score of 98% for accuracy, recall and precision. This was sufficient to predict the occurrence of interfacial precipitates. The precipitate chemistry and morphology were identified by the algorithms as being the most informative parameters for decision making. Disk-shaped particles which contained molybdenum were predicted to be very likely to form interphase precipitates.

Deep convolutional neural networks were used[119] to classify light-microscopy photographs of the microstructures of C15, C45, C60, C80, V33, X70 and carbide-free steels. The neural network-based classification systems required copious training data and the system accuracy depended strongly upon the size of these data-sets. The resultant set (33283) of micrographic images of various types of microstructure provided the opportunity to develop high-precision classification systems; attaining an accuracy of 99.8%.

Artificial intelligence was used[120] to predict the fatigue strength of carbon and low-alloy steel on the basis of composition and processing parameters. A fatigue data-set and deep learning framework were used to build a neural network. The fatigue strengths which were estimated in this way were compared to the values which were predicted by the gradient boosting algorithm. The comparison was performed by using metrics such as the root mean square error, the mean absolute error, the coefficient of determination and the explained variance score. The trained neural network was used to predict of fatigue strength of 10^6 simulated data-points and generate conditional probability tables for a Bayesian network; a system which could exploit both forward- and backward-propagation for Bayesian inference.

A database on compacted-graphite cast iron was used to determine the parameters of a neuro-fuzzy ANFIS model[121]. The database mainly included the results of measurements of the content of microstructural constituents of the iron, as a function of the content of alloying additions such as molybdenum, nickel and copper. Training of the fuzzy inference system was performed by constantly changing its parameters and determining new rules on the basis of individual cases from the training sample. It was shown to be possible to use the ANFIS model to control the chemical composition of compacted graphite iron so as to produce castings of high strength.

A control system for cutting-tool supervision was designed[122] which recognised tool-wear automatically during the dry-turning of the low-alloy carbon steel, 42CrMo4. A data-set of over 9000 images was built up by machining using tool inserts of various wear levels. A convolutional neural network provided a model for tool-wear and tool-damage prediction. The system determined the wear-level (zero, low, medium, high) of a cutting tool automatically on the basis of thermographic processing data. The accuracy of the classification was 99.55%; confirming the adequacy of the method.

A computational approach to estimating tool-wear during turning used support vector machines for regression with Bayesian optimization[123]. A coated-insert carbide tool was used to turn 709M40 alloy steel, and data on feed-rate, depth-of-cut and cutting-speed were collected while tool-wear was monitored via scanning electron microscopy. Other work[124] has incidentally shown that out-of-focus scanning electron microscopic images, arising from improper hardware adjustment and imaging errors, can be corrected by using a deep learning-based re-focusing methods based upon convolutional neural networks of single-scale type, multi-scale type and multi-scale type powered by data-augmentation. This method could re-focus low-quality scanning electron microscopic images. The above networks were tested by using images of martensitic steel and precipitation-hardened alloys both qualitatively and quantitatively. In the present case, the support vector machine model was trained using 162 experimental data-points and the trained model was then used to estimate experimental test data in order to judge the model's performance. The proposed support vector model with Bayesian optimization estimated the tool wear with a mean absolute percentage error of 6.13% and root mean square error of 2.29%.

A low-cost microphone, combined with state-of-the-art machine-learning algorithms, has been used[125] to differentiate various materials and processing regimes in laser powder-bed fusion. Three processing regimes (lack-of-fusion pores, conduction mode, keyhole pores) and 3 alloys (316L stainless steel, CuSn8 bronze, Inconel-718) were selected. Three conventional machine-learning algorithms and a convolutional neural network were chosen to carry out classification, resulting in 5 main findings. It was proved that the acoustic emission features were related to laser-material interaction and not to machine or environmental noises. The processing regimes were classified with an accuracy of better than 87% regardless of the algorithm or material involved. It was possible to build a single model from the 3 materials and maintain a classification accuracy of more than 86% for the various regimes. The acoustic emission features which were used for the classifications were material- and regime-dependent. A strategy for classifying the materials and the processing regimes simultaneously attained a classification accuracy of about 93%.

An attempt was made to analyze, model and optimize the machining of AISI 4140 hardened steel by using self-propelled rotary tools[126]. Experimental testing compared fixed and rotary tool performance and determined the effect of the inclination angle upon the surface quality and tool wear. Genetic programming was used to model and optimize the machining process, with the feed-rate, cutting-velocity and inclination-angle as the inputs and tool-wear, surface roughness and material removal-rate as the outputs. The optimum surface roughness was found at a cutting-speed of 240m/min, an inclination-angle of 20° and a feed-rate of 0.1mm/rev. The minimum tool flank-wear was found at a cutting-speed o[127]f 70m/min, an inclination angle of 10° and a feed-rate of 0.15mm/rev. Differing weights were assigned to the 3 outputs so as to offer various optimized solutions biased in terms of finish or productivity.

The identification of micro-features on 2.25Cr1Mo0.25V steel is important for understanding the mechanism of hydrogen embrittlement, and hence embrittlement-resistance. Transformer-based neural networks perform better than do convolution neural networks with regard to the learning of global information and offer a higher predictive accuracy. A new transformer-based neural network model was proposed in which the architecture of the decoder was re-designed so as to more precisely detect and identify micro-feature having a complex morphology, such as dimples on a steel fracture surface. Various models were compared using a dimple data-set which consisted of 830 scanning electron microscopic dimple images on a steel fracture surface.

Empirical equations have been proposed for predicting austenite grain-growth during re-heating, but it is important to improve the accuracy of prediction models and analyze model mechanisms. Machine-learning models enhance predictive accuracy without requiring additional experimentation. Machine-learning models were used[128] to predict austenite grain-growth with greater accuracy and so-called explainable artificial intelligence was used to treat the problem by collecting 458 data-points from the literature; analyzing and eliminating outliers. Random forest regression was compared with an empirical equation in order to confirm enhancement of the model accuracy.

An active-learning approach was combined with first-principles calculations in order to search rapidly for potentially stable crystal structures, to clarify their stabilization mechanism and to integrate the approach with $SmFe_{12}$-based compounds, having a $ThMn_{12}$ structure, that possesses marked magnetic properties[129]. The aim was to estimate accurately the formation energies via limited first-principles calculation data, to monitor visually the progress of the structure search-process, to deduce correlations between structures and formation energies and to choose the most suitable candidate substituted-$SmFe_{12}$ structures for subsequent first-principles calculations. The structures of $SmFe_{12-\alpha-\beta}X_\alpha Y_\beta$, before optimization, were prepared by substituting X and Y elements such as molybdenum, zinc, cobalt, copper, titanium, aluminium and gallium, within the range of $\alpha+\beta < 4$, into the iron sites of the original $SmFe_{12}$ structures. By using the optimized structures and formation energies which were deduced from the first-principles calculations after each learning cycle, an embedded 2-dimensional space was constructed with which to visualize clearly the scope of all of the calculated and to-be-calculated structures and guide the progress of the search. The machine-learning model had a prediction-error, for the formation-energy, of 1.25×10^{-2} eV/atom. The time required to identify the most potentially stable structures was about one quarter that of a random search. Substitutions of aluminium and gallium had the greatest ability to stabilize the $SmFe_{12}$ structure, and $SmFe_9(Al,Ga)_2Ti$ exhibited the highest stability. Correlations were found between the coordination number of the substitutional sites and the resultant formation energy. Compounds with negative formation-energy of the form, $SmFe_{12-\alpha-\beta}(Al,Ga)_\alpha Y_\beta$, exhibited a common trend of increasing coordination number at substituted sites. Structures with positive formation energy exhibited a decreasing trend.

An investigation was made[130] of liquid-phase formation when a Fe–Ni–Mn–C–B master-alloy was used as a boron-carrier together with 2 iron-based powders which were pre-alloyed with molybdenum. By combining differential scanning calorimetry, artificial intelligence and measurements of the sintered density and strength as a function of sintering temperature, it was possible to clarify the mechanisms which occurred before and during liquid-phase sintering. A series of steps was shown to occur in the solid state so that before reaching the temperature at which a eutectic reaction produced the liquid. The pre-alloyed molybdenum affected the initiation of liquid-phase sintering but not the mechanisms themselves.

On-line mechanical properties were predicted[131], for the tandem hot-rolling of structural carbon and low-alloy high-strength construction steels, on the basis of processing parameters, by using neural network, random forest and other artificial intelligence algorithms. The system offered good precision, high stability and reliability. The predictive accuracy of the model was within ±6%, the sample volume attained more than 90% and the average absolute percentage error was ≤4%; lower than the reproducibility detection level.

Numerical simulations were used[132] to determine the effect of pulse duration and frequency upon the temperature distribution and velocity field during the laser-welding of SS420 stainless steel, SS304 stainless steel and Bohler 303. The results showed that Marangoni and buoyancy forces were the most critical aspects of the flow of liquid metal. An artificial intelligence method could provide optimum predictions of the melting ratio and maximum temperature of the material. A combination of artificial neural networks and particle swarm optimization was used. The latter algorithm optimized the architecture and training of the artificial neural network while the network itself was used for regression. A 3-layer feed-forward architecture, with sigmoidal transfer functions having 17 and 8 neurons in the hidden layers combined with a scaled conjugate gradient back-propagation training scheme was the optimum configuration. Application of the optimum artificial neural network to the regression problem led to an acceptable level of error for the training, validation and testing data-sets.

By using artificial intelligence and machine-learning methods, the time-temperature-transformation curves of steel could be predicted on the basis of limited experimental data[133]. With reliable data as training material, the curve for alloy structural steel could be predicted using various machine-learning algorithms. The alloying elements, austenitizing temperature and phase transformation time were the inputs, and 10 types of transformation characteristic were the outputs. The correlation coefficient and error analysis were used to evaluate the model, and the best algorithm was chosen to form a combined machine-learning algorithm with which to predict the time-temperature-transformation curve. It was applied with success to 40Cr, 38CrMoAl, 35SiMn and 20Mn2 steels.

Artificial intelligence based techniques were used[134] to determine the roles played by composition and processing in deciding the mechanical properties of API-grade micro-alloyed pipeline steel. The aim was to design a steel which offered an improved performance in terms of strength, impact-toughness and ductility. Artificial neural network models were used to deduce the relationships between composition, processing parameters and mechanical properties. The models drove multi-objective genetic algorithms for guiding trade-offs between the conflicting objectives of improved strength, ductility and toughness. Numerous plots, like those of figures 22 and 23 revealed the effect of titanium and vanadium on various mechanical properties. Titanium has a greater effect when compared with those of niobium and vanadium. The latter had a markedly positive effect upon strength, but was detrimental to ductility and impact-energy. It was noted that vanadium carbide or carbonitride have lower dissolution temperature in austenite and were therefore not so effective in producing grain-refinement and the formation of fine acicular ferrite or granular bainite. They could thus be effective in increasing the strength of ferrite via precipitation-hardening, but not effective in effecting grain-refinement nor microstructural changes. The effect of re-heat temperature, in conjunction with the titanium content could also be plotted (figures 24 and 25). In the absence of titanium, the yield-strength decreased up to a re-heat temperature of 1100C.

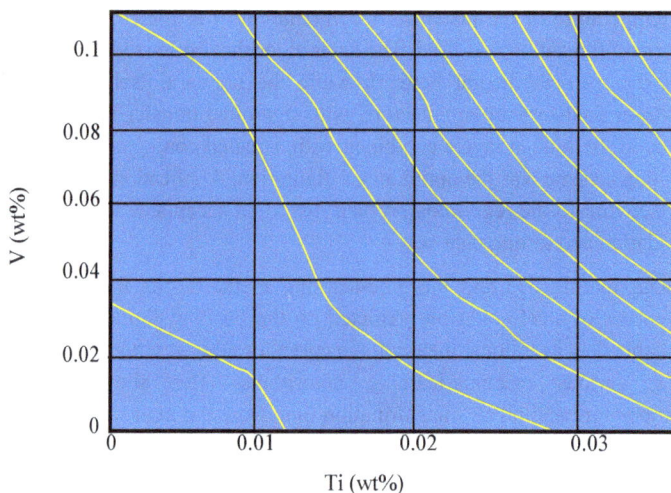

Figure 22. Effect of vanadium and titanium on the UTS of pipeline steel
The contours range from 660 to 840MPa from bottom-left to top-right

At higher temperatures, it continually increased. With increasing titanium content, this transition-temperature gradually decreased to lower temperature. It was noted that Nb(C,N) dissolves in austenite at about 1100C and therefore the yield-strength was tentatively ascribed to that. With increasing re-heat temperature the UTS decreased and, with increasing titanium content, the trend increased towards higher temperatures. This

Materials Research Forum LLC
https://doi.org/10.21741/9781644903148

was possibly due to a higher fraction of stable titanium compounds being formed in these steels and at those re-heat temperatures. The elongation continually increased with the re-heat temperature in the absence of titanium but, with increasing titanium content, the re-heat temperature required for a given strength-level decreased up to between 0.02 and 0.025wt%. The re-heat temperature then again increased. The impact-energy exhibited a similar behaviour to that of the yield-strength but, in this case, the transition temperature fell much faster. A higher cooling-rate was known to produce a higher yield-strength and UTS. In case of elongation, it decreased. In general, an increased cooling-rate led to a better strength but poor formability. The effect of niobium on the yield-strength and UTS was greater than that of vanadium. These elements increased the strength and elongation in different ways. Increasing the niobium content increased the elongation, with a maxima at 0.06wt%, while the elongation decreased with increasing vanadium content.

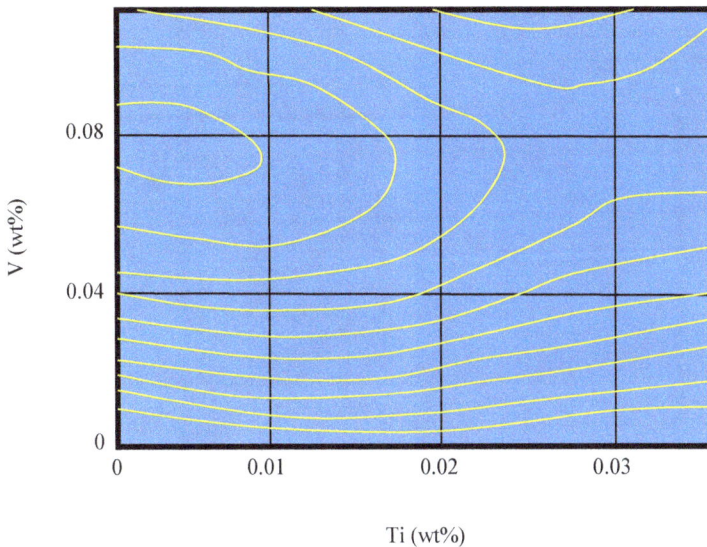

Figure 23. Effect of vanadium and titanium on the impact-energy of pipeline steel. The contours range from 150 to 195J from the centre outwards

Very similar effects occurred in the case of impact-energy, apart from a cusp at a certain niobium content which decreased as the vanadium content decreased. The steel offered a better strength and impact-energy, but poor ductility, upon increasing the content of vanadium. These effects were attributed to the formation of V(C,N) at lower temperatures, as compared to the Nb(C,N) which formed at higher temperatures. A key conclusion was that, for pipeline-grade steel, a composition with niobium and vanadium, but without titanium, was optimum for obtaining the best toughness (400J) at room temperature.

Machine learning was used[135] to estimate the hardness of a traditional spheroidal cast iron and of graphitic cast iron. Microstructures were used as the inputs for training the neural networks, with hardness values as outputs. The training data comprised measurements which characterized metallurgical features such as graphite, ferrite, pearlite, nodularity and vermicularity. The data were in the former of single values estimated from micrographs. Each sample of either cast iron provided a set of 5 values. Each set of metallographic characteristics was combined with the related measured hardness value. The neural network was taught using these data and furnished outputs in terms of Brinell hardness, as estimated using 3 predictive methods: random forest, neural network, k-nearest neighbours (tables 18 and 19).

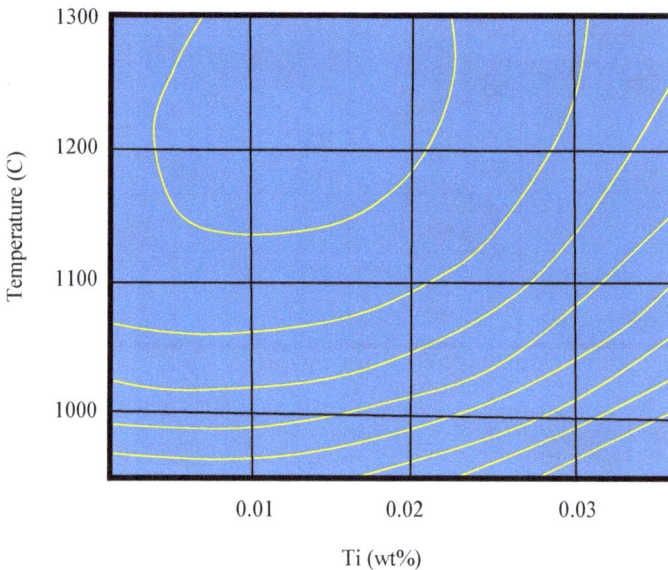

Figure 24. Effect of re-heat temperature and titanium on the UTS of pipeline steel. The contours range from 680 to 820MPa from the centre outwards

The estimates which were provided by the neural network method were considered to be the best, with Pearson correlation coefficients of 0.59 and 0.43 in the case of spheroidal and graphitic cast iron, respectively. It was concluded that open-source self-learning algorithms, combined with databases consisting of just 20 to 30 data-points were already capable of predicting the hardness, with errors of less than 15%. It was expected that greater accuracy would easily be possible increasing the size of the data-set upon which the artificial neural network was trained or by optimizing the neural network in terms of depth and quality of analysis and using other estimation methods, such as multiple regression, nearest-neighbours, genetic programming or support vector machines.

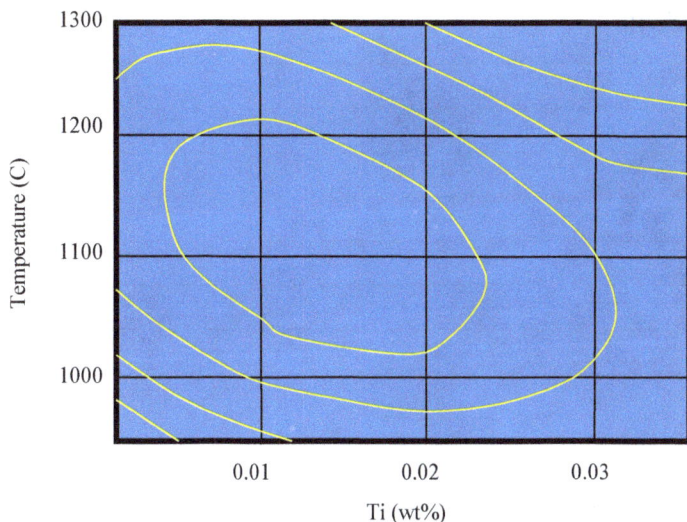

Figure 25. Effect of re-heat temperature and titanium on the impact-energy of pipeline steel.
The contours range from 175 to 190J outwards

Materials Research Forum LLC
https://doi.org/10.21741/9781644903148

Table 18. Measured and predicted hardness values for spheroidal cast iron

Measured(H$_B$)	Random Forest	Neural Network	k-Nearest-Neighbour
165	182	168	181
166	174	171	171
167	178	178	173
168	182	182	169
169	182	171	168
171	182	182	169
171	182	182	166
173	171	182	171
173	182	184	165
174	181	178	167
176	204	184	165
178	182	181	165
178	181	182	171
180	176	206	183
181	178	178	169
181	173	173	165
182	178	178	171
182	173	171	165
182	178	171	171
183	184	206	180
184	180	176	176
185	169	182	169
186	190	204	180
190	185	206	180

204	206	206	206
206	183	190	180
206	204	204	180

Artificial intelligence methods were used[136] to characterise Fe-Zn coatings by using nano-indentation, which could measure the mechanical properties of films thinner than 1µm. The characterization was based upon studying the load-displacement curves of the nano-indentation tests. A back-tracking search optimization algorithm was used to improve the extraction of data from nano-indentation tests of galvannealed Fe-Zn coatings. When applied to coatings on a dual-phase DP600 steel substrate the back-tracking algorithm offered better precision and less dispersion when compared with the least-squares approach. The average errors in the case of the back-tracking algorithm and least-squares methods were 0.89% and 3.16%, respectively, for the yield stress and 3.17% and 7.93% for the strain-hardening exponent. This reduced the error variability in the prediction of the constitutive law to 72% and 60% for the yield stress and strain-hardening exponent, respectively.

Table 19. Measured and predicted hardness values for graphitic cast iron

Measured (H$_B$)	Random Forest	Neural Network	k-Nearest-Neighbour
132	148	137	137
136	141	141	139
137	145	145	142
139	142	136	136
141	142	144	142
142	147	144	141
142	151	156	147
144	142	149	141
144	142	145	137
145	132	137	137

146	150	149	147
147	147	132	132
147	147	151	151
147	147	151	151
148	150	147	132
149	147	156	144
150	146	148	146
150	142	147	142
151	147	147	147
151	147	147	147
156	147	151	141

A Gaussian process regression Bayesian optimization algorithm was used[137] to develop a zirconia-based coating, for Fe-Si alloy sheets, which offered a high heat-resistance and good surface properties. The optimized coating material exhibited a high-quality silvery-white surface, and showed no damage after heat-treatment at above 1100K, with a surface current value of 600mA or less; a measure of insulation. When compared with traditional trial-and-error methods, the number of experiments which was required to achieve the target requirements was reduced to less than 0.1% by using the present artificial intelligence method.

A study was made[138] of the optimization of the metalloid contents of Fe-Si-B-C amorphous soft magnetic materials by means of artificial intelligence methods. The magnetic properties of such alloys can be controlled by varying their composition. Experimental data reveal some departure from the predicted values. Machine-learning methods such as random-forest regression, k-nearest neighbours regression and support-vector regression can help to optimize the composition. As a result, the composition possessing the lowest coercivity was found to be $Fe_{80.5}Si_{3.63}B_{13.54}C_{2.33}$ and the composition possessing the highest saturation magnetization was $Fe_{81.83}Si_{3.63}B_{12.63}C_{1.91}$. The R^2 values were 0.74 and 0.878, respectively.

The Shaply additive explanation value offered an intuitive understanding of the effect of relevant variables upon creep-life[139]. The artificial intelligence model was optimized by using a genetic algorithm, and led to the identification of an optimum alloy composition with a required creep-life could be presented. The longest creep-life which corresponded to the composition and heat-treatment of ASMESA213-T92 was deduced by using a genetic algorithm method. The maximum creep life was predicted to be 58640 hours. On the basis of previous knowledge, sulphur was treated as a harmful element because it reduced creep ductility. On the other hand, sulphur additions were predicted to increase creep-life. This demonstrated that alloys must be designed with consideration of various requirements. According to the genetic algorithm predictions, it was advisable that the silicon, nickel and aluminium contents be as low as possible. In the case of tungsten, it was predicted that an appreciable creep-life increase would start to appear from 1.8wt% and, for boron, from 0.003wt%. Molybdenum was not greatly increased significantly, but the creep-life continued to increase. In the case of these alloy components, the maximum life was at the maximum allowable composition. The maximum creep-life was predicted to be at 0.23%V, at 0.09wt%C and at 9.10%Cr. The lifespan could be rapidly reduced at 8.6%Cr or less. The creep-life was improved by about 40000 hours upon adding 2.5%Co. This tendency was found for other alloying elements. Cobalt did not have a clear interaction with other alloying elements, and it was determined that it strongly contributed to the solid-solution strengthening effect. As the cobalt content increased, the creep-life increased up to a maximum of 114800 hours. It was surmised that it would be possible to develop a new alloy composition which would result in a creep-life of 100000 hours at 650C under a load of 100MPa. Among the artificial intelligence models, XGB was optimum in an ensemble which comprised deep nearest-neighbour, gradient boost and light gradient boost models. Based upon the extra gradient boost model, it was shown that the creep-life was affected – in order - by the normalization temperature, and by the contents of vanadium, tungsten, molybdenum and carbon. The current life-prediction model could not in fact suggest an alloy composition having a required creep-life but, by using the genetic algorithm method, an optimum alloy composition could be obtained. By adding cobalt to the composition of ASMESA213-T92, a new alloy with a life of 100000 hours at 650C under a load of 100MPa was predicted to have the composition: 0.09C, 0.03Si, 0.03Mn, 0.31B, 0.015P, 0.01S, 9.10Cr, 0.59Mo, 1.99W, 0.01Ni, 0.23V, 07Nb, 0.06N, 0.001Al, 0.006B, 2.5Co.

Magnesium Alloys

The room-temperature ductility of magnesium alloys was closely related to the deformation behaviour of twins, but there were no criteria that could accurately predict those grains within which twins would nucleate during plastic deformation[140]. A Mg-4Y-3Nd-2Sm-0.5Zr alloy was prepared so as to investigate the twin nucleation behaviour by combining machine-learning and electron back-scattered diffraction data. At a true strain of 0.05, twins were found in 68 grains from among 297 grains selected in the initial microstructure. Eight factors were chosen that might affect twin nucleation. These included the grain diameter, the number of neighbouring grains and the Schmid factor. The original diameter of the grain and the Schmid factor of tensile twins had the greatest influence upon twin nucleation. Three machine-learning algorithms were used: XGBoost, artificial neural network and a proposed relevance-based ensemble scheme. The latter was superior to the others.

An attempt was made[141] to find the optimum combination of parameters, such as die-temperature, step-distance, lower punch radius, die clearance and step-height which would minimise springback; the undesirable phenomenon which often occurs during sheet-metal forming. The combination of a finite-element model and Taguchi experimental design was used to form a design matrix. An adaptive neuro-fuzzy inference system was then used to correlate relationships between the process inputs and springback. The taught machine-learning algorithm was combined with the ANFIS model so as to minimize springback. The proposed approach could accurately predict the optimum drawing process.

A machine-learning model was able to relate the grain-size, tensile yield stress, compressive yield stress, ultimate tensile strength, ultimate compressive strength, compressive and tensile strain-to-failure, hardness and texture of indirectly-extruded Mg-Gd alloys to processing parameters such as the extrusion velocity and temperature[142]. The gadolinium content was between 0 and 10%. Shallow artificial neural networks were used to predict the material properties. It was possible to predict the anisotropic behaviour of Mg-Gd alloy. The machine-learning model was slightly better in predicting the material properties, as compared with a linear-regression approach.

The identification of regions which were likely to exhibit twin interaction with grain boundaries in AZ31 alloy was investigated[143] by feeding observations in the neighbourhood of twin activity into a machine-learning framework. After applying an attribute-selection filter and other machine-learning tools, a decision-tree model was able to classify likely neighbourhoods of twin activity with 85% accuracy. An investigation was made[144] of the mechanical and microstructural behaviour of the friction stir-welded AZ31B alloy. An artificial neural network model was used to produce a relationship between processing variables such as the tool rotational speed, the welding speed and the tool shoulder diameter, and characteristics of friction stir-welded joints, such as the tensile strength, the elongation, the impact strength, the microhardness and the grain size. The best optimum solution for a maximum tensile strength of 164.2MPa, an elongation of

8%, an impact strength of 3.5J, a microhardness of $85H_V$ and a minimum grain-size of 13.1μm was confirmed with a less than 3% absolute error.

Soft computing techniques were used[145] to predict the surface integrity of ball-burnished ZE41A alloy. The ANFIS model surpassed other models in estimating the surface roughness with a lowest mean error of 2.20%. The artificial neural network surpassed other two models in estimating the microhardness, with a lowest mean error of 0.37%. The mean error of all the soft computing models was less than 5%. Nano-ZrO_2 particles were incorporated[146] into cast rare-earth ZE41 alloy by means of friction stir processing and a surface composite layer of $ZE41/ZrO_2$ was produced. The relationship between the friction stir processing parameters, grain-size and hardness was determined by using multiple regression and artificial neural network techniques. The ZrO_2 particles were not uniformly distributed within the stir zone, and the zone was divided into 2 regions. The model input parameters were the tool rotational speed, the tool traverse speed and the type of stir-zone region. The outputs were the grain-size and hardness. The models could then be used to predict the grain-size and the hardness as a function of rotational speed, traverse speed and type of region. The mean square error for grain-size was 0.3418 (regression model) and 0.000121 (neural network), while the mean square error for hardness was 1.0211 (regression model) and 0.00018 (neural network). The coefficient-of-correlation, R^2, of the training and testing for grain-size and hardness was close to unity.

Six supervised machine-learning regression-based algorithms, decision tree, XGBoost, artificial neural networks, random forest, gradient boosting, and AdaBoost were used[147] to predict the ultimate tensile strength of the friction stir welded magnesium joints. The alloys tested were AM20, AZ61A, AZ31B and AZ31. The plunge-depth, the shoulder-diameter, the tool traverse-speed, the pin-diameter, the axial force and the tool rotational speed were the input parameters while the ultimate tensile strength of the welded joint was the output parameter. It was found that the alloy-type had the greatest influence among the input parameters. The XGBoost algorithm offered the highest coefficient of determination (0.81).

Nickel Alloys

Spindle-power data for real-time tool-wear and tool-breakage monitoring during the drilling of Inconel-625 were obtained[148] for various speeds and feed-rates and fed to a neural network for processing. Force data were also collected and processed. The results showed that the trends exhibited by the two types of data were similar for any combinations of feed-rate and speed. It was concluded that spindle-power data, combined with an artificial intelligence system could be used for real-time tool-wear and breakage monitoring.

Genetic algorithms and particle swarm optimization were used[149] to train artificial neural networks for predicting the cutting-forces involved in the turning of Inconel-718, as assisted by high-pressure coolant. These results were compared with those of commonly used back-propagation based artificial neural networks, showing that the training of neural networks by using bio-inspired algorithms gave better results. Hybrid machine-learning models were developed for the prediction of the induced residual stresses created during the turning of Inconel-718 alloy[150]. These models comprised a traditional artificial neural network and bio-inspired optimizers: a pigeon optimization algorithm and a particle swarm optimization algorithm. These were used to fine-tune the neural network parameters so as to increase its predictive accuracy. The models were trained by using residual stress data measured under various cutting conditions. The predicted residual stresses were compared with measured ones as well as with those predicted by traditional artificial neural networks. The predictive accuracy of the models was evaluated by using 7 statistical parameters. The pigeon and particle swarm hybrids out-performed the traditional artificial neural network. The coefficient-of-determination of the pigeon optimisation, the particle optimisation and the traditional neural network was 0.991, 0.938 and 0.585, respectively, and the root mean square errors were 11.870, 31.487 and 119.437, respectively. A hybrid optimization artificial intelligence tool was designed for the selection of the best processing parameters for obtaining desired properties[151]. Tests were designed by using the Taguchi technique and considering factors such as the laser power, the weld speed and the pulse duration. Experiments on thin Inconel-718 sheets were used to optimise the required output measurements. The process parameters were optimised by using the Grey-ANFIS-based Jaya algorithm to achieve the smallest top/bottom width and greatest penetration. Grey theory was used to establish a multi-performance index, and the performance of the optimization model was assessed using statistical error analysis. A predictive model for metal flow during hot deformation was tested[152] using Inconel-718, an alloy which is rather intractable and requires care in optimizing its deformation, microstructure and properties. A rheological model was developed by using an adaptive neuro-fuzzy inference system and data gleaned from compression testing using a Gleeble 3800 thermo-mechanical simulator at 900, 1000, 1050, 1100 and 1150C at strain-rates of 0.01/s to 100/s and a constant true strain of 0.9. The values of yield stress which were predicted by the model were favourably compared with experimental results.

Machine-learning was used to predict[153] the lattice misfit, between γ and γ' phases, in nickel-based monocrystalline superalloys which critically affects microstructural stability and high-temperature creep and fatigue resistance. The inputs included the chemical composition, dendrite parameters and the temperature. Support vector regression, sequential minimal optimization regression and multilayer perceptron algorithms plus experimental data gave predictions with high correlation coefficients and low error values.

Nickel-based alloys were tested[154] using 1h cycles at 800 to 950C in wet air, and the oxide scale which formed on wrought Ni-(14-25)wt%Cr binary alloys was used to monitor the behaviour of chromia-forming alloys. Mass-change curves were used to quantify the behaviour of the tested alloys and to fit growth and spallation rates. A systematic analysis of the correlation between alloy composition and oxidation data was used to select the most important features to be included in a machine-learning model. Machine-learning models for the corrosion rate could be trained with a surprisingly high degree of accuracy, even in the presence of limited data. On the other hand, only moderately good fitting was obtained for the spallation parameter. A theoretical model which could predict the corrosion and spallation rates of hypothetical alloys was developed.

An artificial intelligence technique was used to optimise the precipitation-hardening of a nickel-based alloy so as to allow more flexible non-isothermal aging[155]. A computational method was used to model the microstructural evolution and evaluate the 0.2% proof stress for isothermal aging and non-isothermal aging. It was possible to obtain an improved 0.2% proof stress for non-isothermal aging at a fixed time of 600s, as compared with isothermal aging. Among 1620 non-isothermal aging schedules, 110 schedules were designed which out-performed the isothermal aging equivalent. Early-stage high-temperature aging for a shorter period increased the γ' precipitate size up to the critical value. Later aging at a lower temperature increased the γ' fraction, with no anomalous change in the γ' size.

An accurate machine-learning regression model for the configurational energy of a multicomponent solid solution could be constructed by training using density functional theory calculations[156]. A feature vector, which was formed by combining correlation functions, encoded the spatial arrangement of atoms in the lattice and provided a good description of the configuration. The feature vector concatenated histograms of pair and triplet correlation functions which constituted a quantitative fingerprint of the spatial arrangement of atoms. The predictive accuracy of the machine-learning model was comparable to that of the cluster expansion approach. The model could be used to generate a distribution of configurational energies by rapidly sampling the configurational space. The (100) γ/γ' interface energy (table 20) ($25.95mJ/m^2$) at 700C in

5-component alloy 617, a nickel-based superalloy, when calculated by using the machine-learning model exhibited good agreement with the fitting of experimental precipitation data. The model could potentially be used to calculate the parameters and properties of alloys having any number of components.

Table 20. Formation energies of γ/γ' interfaces in nickel alloys

$E_{\gamma/\gamma'}$(eV)*	Misfit(%)	γ Structure	γ' Structure	Formation Energy(kJ/mol/atom)
−816.218	1.182	L1	L1	1.273
−816.049	1.177	L1	L2	1.446
−815.952	1.138	L1	M1	1.206
−816.120	1.090	L1	M2	0.966
−815.725	1.201	L1	H1	0.594
−814.994	1.198	L1	H2	1.143
−816.413	1.196	L2	L1	0.943
−815.589	1.190	L2	L2	1.609
−816.102	1.151	L2	M1	0.909
−815.834	1.103	L2	M2	0.998
−815.704	1.214	L2	H1	0.426
−815.040	1.211	L2	H2	0.925
−816.270	0.784	M1	L1	0.186
−816.023	0.779	M1	L2	0.418
−815.096	0.740	M1	M1	0.803
−815.388	0.692	M1	M2	0.470
−815.183	0.803	M1	H1	−0.045
−814.237	0.800	M1	H2	0.666
−815.717	0.960	M2	L1	0.416
−815.489	0.955	M2	L2	0.634

−815.533	0.916	M2	M1	0.287
−815.609	0.868	M2	M2	0.116
−815.316	0.979	M2	H1	−0.333
−814.545	0.976	M2	H2	0.246
−814.055	0.785	H1	L1	0.991
−813.803	0.780	H1	L2	1.227
−813.855	0.741	H1	M1	0.874
−813.803	0.693	H1	M2	0.800
−813.910	0.804	H1	H1	0.049
−812.838	0.801	H1	H2	0.856
−815.044	0.933	H2	L1	0.123
−814.920	0.928	H2	L2	0.263
−814.939	0.889	H2	M1	−0.065
−814.600	0.841	H2	M2	0.078
−814.347	0.952	H2	H1	−0.402
−813.838	0.949	H2	H2	−0.020

*Fully relaxed energy of the γ/γ′ structure.

A ternary Ni-Ti-Hf shape memory alloy with an austenite finish temperature above 400C was designed[157]. Available experimental data on the ternary system were used to compile a database for training and testing a machine-learning algorithm so as to predict the ideal composition with the desired finish temperature plus a relatively small hysteresis. A multi-layer feed-forward neural network model was trained and tested. The $Ni_{49.7}Ti_{26.6}Hf_{23.7}$ and $Ni_{50}Ti_{27}Hf_{23}$ alloys which were predicted by the machine-learning algorithm were investigated in order to assess the accuracy of the model predictions. The former alloy, with an austenite finish temperature of 403.5C and marked cyclic stability, was chosen as a new alloy for applications which might require a reversible austenite-to-martensite phase transformation beyond 400C.

High-throughput first-principles calculations were used[158] to predict the synergistic effect of aluminium, plus one of 28 choices of 3d, 4d or 5d transition metals, upon the elastic constants of γ-nickel. Machine-learning methods predicted the relationship between alloying addition and mechanical properties upon being fed with the binding energy between the 2 elements. It was discovered that ternary elements strengthened the γ-phase in the order: Re > Ir >W> Ru > Cr > Mo > Pt > Ta > Co, and that there was a parabolic relationship between the number of d-shell electrons of the alloying element and the bulk modulus. The maximum bulk modulus occurred when the d-shell was half-full. There was a linear relationship between the bulk modulus and the alloy concentration over certain ranges. Linear regression revealed a linear fit for the concentrations of 29 elements. For the bulk modulus and lattice constants of $Ni_{32}XY$, the predicted values were close to calculated results; with a regression parameter, R^2, of 0.99626. As compared with pure nickel, the nickel alloys had higher B, G, E, C_{11} and C_{44} values but equal C_{12} values. Changing the size of the unit cell changed the concentration of the alloying element. It was possible to predict successfully the behaviour of 28 Ni_3X_2 and 435 Ni_3XY systems. An R^2 value of 0.99431 confirmed the accuracy of the predicted relationship between the modulus and concentration. Analysis of the dependence of the bulk modulus, elastic constants, elastic moduli and Poisson ratio for 435 Ni_3XY ternary systems showed that the addition of an element which increased the bulk modulus would also increase the shear modulus, the Young's modulus, C_{11} and C_{44} without decreasing the other parameters. In general, the choice of a given alloying elements affected more than one parameter. Those elements which increased the bulk moduli of all of these systems also reduced the system binding-energy (figure 26), thus indicating that the latter was a good criterion for assessing the elastic properties of an alloy. Nickel-based alloys exhibited a negative correlation between the binding-energy of the alloying element in solid-solution and the elastic modulus. There was a linear relationship between the lattice-constant (tables 21 and 22) and the alloy concentration, with an R^2 value of 0.99837. From the differing lattice-constants of the elements in the two phases, it was possible to predict the mismatch of the alloy.

Table 21. Elastic constants of $Ni_{31}X$ systems

X	B(GPa)	C$_{11}$(GPa)	C$_{12}$(GPa)	C$_{44}$(GPa)	G(GPa)	E(GPa)
Ag	190.63	238.04	166.92	106.00	68.49	183.49
Al	194.04	248.40	166.86	114.32	75.65	200.85
Au	193.43	241.49	169.41	106.07	68.87	184.69

Cd	188.43	236.67	164.32	107.51	69.54	185.77
Co	197.82	250.68	171.39	112.60	74.14	197.71
Cr	198.98	256.84	170.04	121.21	80.34	212.42
Cu	194.34	244.91	169.05	111.96	72.61	193.70
Fe	198.69	254.27	170.89	116.29	77.11	204.84
Hf	192.07	242.61	166.81	107.33	70.75	189.04
Ir	200.43	255.50	172.89	116.05	76.74	204.16
Mn	199.09	256.17	170.55	117.80	78.53	208.22
Mo	198.95	254.20	171.32	116.01	76.82	204.18
Nb	196.04	248.80	169.66	111.49	73.64	196.34
Ni	196.41	246.88	171.17	109.82	71.71	191.79
Os	201.51	259.13	172.70	120.71	80.00	211.96
Pd	194.84	244.13	170.19	107.28	70.05	187.65
Pt	197.88	247.84	172.90	104.83	69.43	186.49
Re	201.39	258.73	172.72	119.90	79.52	210.82
Rh	197.79	250.70	171.34	112.25	74.02	197.44
Ru	199.40	255.21	171.50	116.55	77.34	205.46
Sc	187.38	236.51	162.81	102.23	67.94	181.83
Ta	197.39	250.82	170.68	113.32	74.74	199.09
Tc	199.80	256.05	171.67	116.49	77.56	206.01
Ti	194.76	247.80	168.25	110.57	73.42	195.67
V	197.63	253.47	169.71	117.38	77.69	206.06
W	200.25	255.83	172.46	116.88	77.35	205.58
Y	180.01	225.39	157.32	95.38	63.13	169.57
Zn	192.96	244.85	167.02	110.66	72.83	194.06
Zr	190.32	239.64	165.66	104.53	68.96	184.58

Flow-stress predictions were performed[159] by using conventional constitutive equations and artificial intelligence techniques (figure 27). It was shown that the flow stress, as modelled by using artificial neural networks offered greater accuracy than did the flow stress as modelled by using the conventional Arrhenius hyperbolic sine equation. Mathematical modelling of the flow stress, as based upon the Arrhenius constitutive equation suffered from limitations in flow-stress prediction at high strain- rates. An artificial neural network was used instead. It was applied to 54 cases (table 23) in which the numbers of layers and nodes were varied. There was no significant dependence upon the number of hidden layers but, when 12 hidden layers were used, both the absolute average relative error and the absolute error sharply increased. A lower accuracy resulted when 1 to 3 hidden layers were used.

Table 22. Elemental radius and lattice constant of $Ni_{107}X$ and $Ni_{31}X$ alloys

X	r(Å)	$Ni_{107}X$(Å)	$Ni_{31}X$(Å)
Ag	1.44	3.523	3.541
Al	1.43	3.518	3.523
Au	1.44	3.524	3.543
Cd	1.51	3.525	3.547
Co	1.25	3.516	3.515
Cr	1.28	3.516	3.517
Cu	1.28	3.517	3.520
Fe	1.26	3.516	3.514
Hf	1.59	3.525	3.547
Ir	1.355	3.520	3.531
Mn	1.27	3.516	3.515
Mo	1.39	3.521	3.532
Nb	1.46	3.523	3.541
Ni	1.24	3.517	3.517
Os	1.35	3.520	3.529

Pd	1.37	3.521	3.534
Pt	1.385	3.522	3.535
Re	1.37	3.520	3.530
Rh	1.34	3.520	3.530
Ru	1.34	3.520	3.528
Sc	1.62	3.524	3.543
Ta	1.46	3.523	3.539
Tc	1.36	3.520	3.529
Ti	1.47	3.520	3.530
V	1.34	3.518	3.521
W	1.39	3.521	3.533
Y	1.8	3.531	3.565
Zn	1.34	3.519	3.524
Zr	1.6	3.526	3.551

It was suggested that this was because too many hidden layers caused an over-fitting problem. It was proposed that 6 hidden layers was the optimum number of layers, as judged by the absolute error results. The number of nodes nevertheless had a large effect upon predictive ability. The smallest lowest absolute average relative errors and the absolute errors were measured when using 11 nodes. The best choice was therefore to use 6 hidden layers and 11 nodes when modelling the relationship between deformation and flow stress.

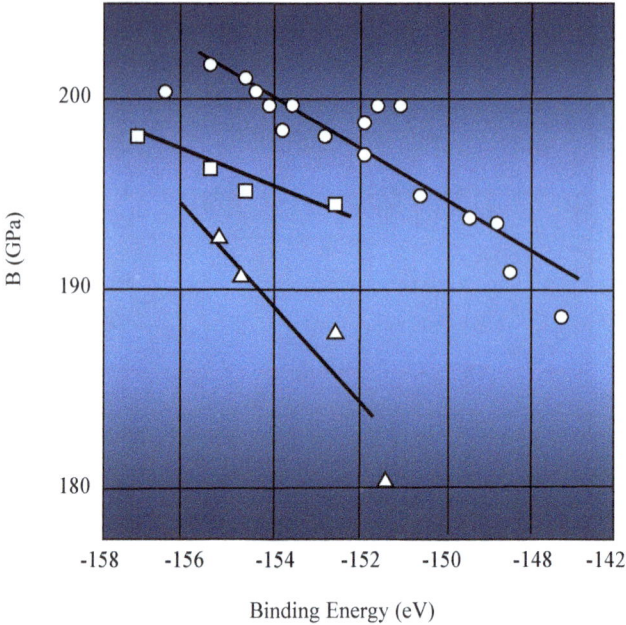

Figure 26. Bulk modulus as a function of binding energy for Ni₃₁X. Elements (X) which form a solid solution in the γ phase of the high-temperature alloy (e.g. Re, Mo, Ru, Co, Cr) exhibit a negative linear relationship (upper lined) and Fe, Cr, Co and Mn deviate slightly from that line. Elements which form a γ' phase (Al, Ta, Nb, Ti) exhibit a less negative linear relationship (middle line). The elements Hf, Zr and Sc exhibit the most negative linear relationship (lower line).

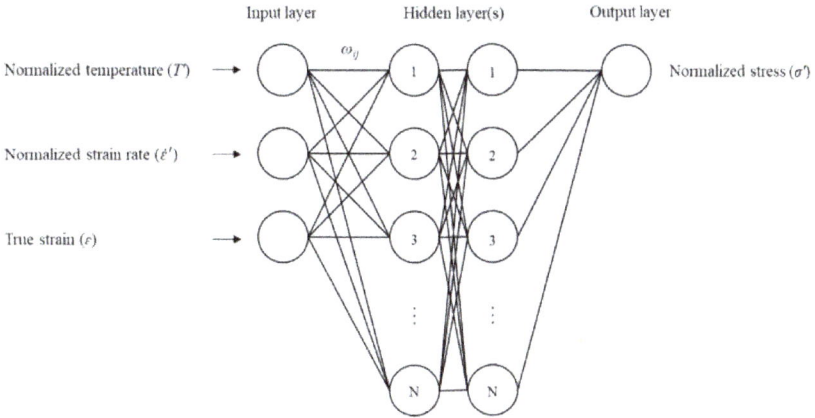

Figure 27. Architecture of the artificial neural network used to predict flow-stress curves. Reproduced from: Predicting high temperature flow stress of nickel alloy A230 based on an artificial neural network, I.Y.Moon, H.W.Jeong, H.W.Lee, S.J.Kim, Y.S.Oh, J.Jung, S.Oh, S.H.Kang, Metals, 12[2] 2022, 223 under Creative Commons licence.

A back-propagation neural network model was used[160] to predict the creep curves of MarM247LC superalloy under various conditions. The prediction errors for the creep curves were within ±20% after using 6 creep-curves to train the model. As compared with the θ projection model, the maximum error was reduced by 30%. The method was also applicable to the prediction of creep curves for superalloys such as DZ125 and CMSX-4.

Table 23. Structures of artificial neural networks used for predictions

Layers	Nodes	AE(MPa)	AARE(%)
4	3	5.0	9.0
4	5	2.9	4.6
4	7	2.5	4.2
4	9	2.3	3.6

4	11	1.3	2.4
4	13	1.5	2.5
5	3	5.7	10.0
5	5	2.7	5.2
5	7	1.9	3.1
5	9	2.0	4.0
5	11	1.3	2.9
5	13	2.2	5.0
6	3	3.1	5.8
6	5	3.4	7.3
6	7	1.8	3.6
6	9	1.8	2.9
6	11	1.2	2.0
6	13	1.2	2.0
7	3	4.5	11.7
7	5	3.3	5.2
7	7	2.8	4.3
7	9	1.6	2.8
7	11	2.0	3.4
7	13	1.9	3.5
8	3	3.0	4.7
8	5	3.0	5.3
8	7	2.3	3.4
8	9	1.6	2.3
8	11	1.6	2.0
8	13	1.7	2.6
9	3	5.1	9.1

9	5	2.7	5.9
9	7	1.9	3.9
9	9	1.4	2.1
9	11	1.4	2.4
9	13	1.6	3.2
10	3	5.2	10.1
10	5	3.7	7.1
10	7	2.0	3.5
10	9	1.6	3.2
10	11	1.5	2.6
10	13	1.1	2.6
11	3	4.1	6.3
11	5	2.8	4.5
11	7	2.1	3.4
11	9	1.9	3.2
11	11	2.0	4.6
11	13	1.2	2.1
12	3	5.1	10.4
12	5	2.8	4.2
12	7	2.5	4.4
12	9	2.6	5.3
12	11	2.8	5.2
12	13	9.6	15.4

AE: absolute error, AARE: absolute average relative error

Titanium Alloys

A study was made of the Ti-Nb-Zr-Sn system in order to identify new composition and temperature ranges which would maximize the stability of the β-phase while minimizing the formation of α" and ω-phase[161]. The CALPHAD (calculation of phase diagram) approach was used to determine the stabilities of the various phases, and those data were analyzed using artificial intelligence algorithms. Deep-learning neural network models were developed for the various phases as a function of alloy composition and temperature. The models were used to predict various phases for new compositions and temperatures. This data-set was further analyzed using self-organizing maps to find correlations between the stability of various phases, the chemical composition and the temperature. Candidate alloys were identified.

A systematic analysis of the effects of hot-deformation processing parameters upon the microstructure of Ti-15-3 alloy was carried out[162]. On the basis of the results, a multi-objective optimization model was established in which the temperature, strain and strain-rate were treated as design variables, with the objective being to obtain a uniform fine-grained microstructure using smaller loads. Optimization of the hot-deformation process was achieved by using an artificial neural network models and a modified genetic algorithm. The latter was effective.

Hot-compression of Ti-6Al-4V was carried out[163] at 1023 to 1323K, using strain-rates of 0.01 to 10/s, by means of a computer-controlled Gleeble-3500 machine. In order to monitor the highly non-linear flow behaviour, support vector regression was combined with a genetic algorithm. The key characteristic of this method was that, given identical training parameters, it would maintain training-accuracy and prediction-accuracy at a stable level during sequential uses of the same data-set. A comparison showed that the learning ability of the present method was better than that of the mathematical regression model, in the modelling-efficiency order: mathematical regression model < artificial neural network < hybrid system. Stress-strain data beyond the experimental conditions were predicted by the well-trained hybrid.

The mechanical behaviour of α+β dual-phase Ti-6Al-4V was studied[164] at 25 to 200C using strain-rates ranging from 10^{-3}/s to 10^{4}/s. Tensile testing was used for the quasi-static range of deformations and the Split-Hopkinson bar was used for the dynamic range of deformation. A modified Johnson-Cook model was developed for the alloy. An artificial neural network framework was used to predict the flow stress for various deformation conditions. It was found that the artificial neural network was more efficient in calculating the flow stress, as compared with the Johnson-Cook model.

At high temperatures titanium alloys become chemically active and tend to react with tool materials. Fuzzy logic was used[165] early on to predict cutting parameters for the turning of Ti-6Al-4V. The parameters which were considered were the cutting speed, the feed-rate and the depth of cut. The fuzzy rule-based modelling could predict the tool-flank wear, the surface roughness and the cutting pressure when machining titanium alloys.

Optimization techniques based upon swarm intelligence, the firefly algorithm, particle swarm optimization and the so-called bat algorithm, were used[166] to optimize machining parameters such as the cutting-speed, feed-rate, depth-of-cut, tool-flank wear and tool-vibrations so as to ensure minimum surface roughness. The performance of the bat algorithm was compared with that of the firefly algorithm and particle swarm optimization, and it was concluded that the bat algorithm offered the best optimisation.

Turning of titanium alloys was optimized with respect to surface roughness, while reducing the effect upon material removal-rate, by using a non-traditional optimization algorithm[167]. Such an optimization algorithm requires a predictive model which identifies an empirical relationship between the parameters of the machining operation and various response-variables. A predictive model was applied to a Ti-6Al-4V turning data-set and used to determine the best technique. An artificial neural network out-performed other predictive models by achieving a mean absolute percentage error of 1.08%, and could handle non-linear and interactive effects in the data. The artificial neural network was then combined with a genetic algorithm so as to offer the best set of parameters for minimizing surface roughness while maximizing the material removal-rate.

An artificial intelligence method was used to generate virtual surface morphologies for Ti-6Al-4V parts, as produced by given processing parameters. A high-resolution surface morphology image-generation system was developed[168] by optimising generative networks. The resultant virtual surfaces well-matched experimental results, with a Fréchet inception distance score of 174 within the range of accurate matching. Microstructural analysis of parts fabricated with artificial-intelligence guidance exhibited a less-textured microstructural behaviour on the surface. The artificial-intelligence guidance of virtual surface morphology could aid the production of high-quality parts.

An investigation was made of the surface characteristics of Ti-6Al-4V selectively laser-melted parts by using image texture parameters[169]. Parameters extracted from surface images by using first-order and second-order statistical methods, and measured 3-dimensional surface-roughness parameters were used to characterize the laser-melted surfaces. Gaussian process regression furnished an accurate prediction of the roughness values, with an R^2 value of better than 0.9.

Tool conditions were tracked by monitoring[170] vibrational and acoustic-emission signatures during the high-speed machining of Ti-6Al-4V. Discrete wavelet transform coefficients of the signals were extracted. Machine-learning algorithms such as decision-tree, naive Bayes, support vector machine and artificial neural networks were used to predict the tool condition. The results showed the effectiveness of using acoustic emission and vibrational data, as wavelets, for classifying tool conditions with the aid of machine-learning. A correlation was identified between the tool conditions and the sensor data. A support vector machine, trained using vibration data appeared to be able to predict the tool conditions with good accuracy, as compared with decision trees, naive Bayes and artificial neural networks.

Heat generation and heat removal from the cutting zone are critical factors in the machining of titanium alloys. The removal process is difficult because of the poor thermal conductivity of titanium alloys, and this leads to rapid tool wear and a poor surface finish. A high-speed precision mill and optimum cutting conditions were used[171] to maximize the life of tungsten carbide tools. Tool wear was judged on the basis of surface roughness. Acoustic-emission signals were monitored by using a sensor during machining. Statistical features of the time and frequency domains were extracted, and features which contained useful information concerning the tool condition were selected by using a decision-tree algorithm.

The surface roughness following the turning of Ti 6Al 4V was predicted[172] by using neural and maximum sensitivity networks. Machining tests were carried out using TiAlN-coated carbide inserts and various cutting conditions. The important parameters were the cutting speed, feed-rate and depth-of-cut.

The micro-machining of Ti-6Al-4V was analysed[173] in terms of the surface roughness of machined microchannels, the cutting-force in the feed-direction, the cross-feed direction and the vibration on the top surface plate along the spindle-axis. Tests were performed using various rotational speeds, feed-per-tooth rates, axial depths-of-cut and cutter-diameters. Significant factors were identified by analysing the variance and the response plots. A rotational speed of 15000rpm, a feed-rate of 0.2μm/tooth, an axial depth-of-cut of 30μm and a cutter diameter of 200 were the optimum settings for ensuring better machinability of Ti-6Al-4V. The analysis of variance results indicated that the feed-rate is the most influential factor in ensuring lower cutting-forces, surface roughness and workpiece-vibration. The stiffness of the micro end-mills greatly affected the vibration amplitude during micro-milling.

In order to improve the micro-end milling of Ti-6Al-4V, a method was proposed for selecting the optimum process parameters which could meet the micro-machining

Artificial Intelligence and Alloy Design   Materials Research Forum LLC
Materials Research Foundations **166** (2024)   https://doi.org/10.21741/9781644903148

requirements and constraints[174]. The experiments, finite-element simulations, constrained multi-objective particle-swarm optimization and mathematical modelling techniques were used to facilitate process parameter selection. Based upon machining tests of a circular thin rib feature, the results indicated a significant improvement in performance. An adaptive neuro-fuzzy inference system was used[175] to predict the surface roughness when end-milling Ti-6Al-4V alloy with coated or uncoated cutting tools under dry-cutting conditions. Experimental results were used to train and test adaptive neuro-fuzzy inference system (ANFIS) models, and the best model was selected on the basis of the minimum root mean square error[176]. A generalized bell-shaped function was chosen as a membership function for the modelling process, and the numbers were changed from 2 to 5. The results were evidence of the ability of ANFIS to model the surface roughness in end-milling and provide a good match between experiment and prediction.

A hybrid adaptive neuro-fuzzy inference system, with a multi-objective particle swarm optimization method, was used to determine[177] the optimum combination of milling parameters and reinforcement ratio that would minimize the feed-force, depth-force and surface roughness. Nanocomposites of Ti-6Al-4V, reinforced with 0, 0.6 or 1.2wt% of graphene nanoplatelets were prepared. A full factorial approach was used to design experiments so as to investigate the effect of the cutting-speed, feed-rate and graphene nanoplatelet ratio upon the machining behaviour. An artificial intelligence method, based upon ANFIS, was then used to develop predictive models. The models could provide accurate estimates of the depth-force, the feed-force and surface roughness; with mean absolute percentage errors of 3.87, 8.56 and 2.21%, respectively, as compared with experimentally measured outputs. The artificial intelligence models also exhibited 361.24, 35.05 and 276.47% lower errors in the depth-force, feed-force and surface roughness, respectively, when compared with traditional mathematical models. The results indicated that a cutting-speed of 62m/min, a feed-rate of 139mm/min and a grapheme nanoplatelet reinforcement ratio of 1.145wt% lead to improved machining characteristics.

In order to monitor tool conditions during the dry-turning of Ti-6Al-4V, a machine-learning method was based[178] upon processing cutting-force, acoustic emission and vibration sensor-signals during turning. In order to reduce the large dimensionality of the sensor features, a methodology which was based upon principal component analysis was used. This permitted the consideration of a smaller number of features, obtained through linear projection of the original d features into a new space of reduced dimensionality. By feeding artificial neural networks with the principal component analysis features, an accurate diagnosis of tool flank-wear was possible, with its predicted values being very close to measured tool wear values.

Predictions of machining-induced microhardness and grain-size were made[179] by using 3-dimensional finite-element simulations of machining, and machine-learning models, in the case of the machined surfaces of Ti-6Al-4V. Hardness measurements were made at high temperatures in order to develop a predictive model by utilizing finite-element based temperature fields for the hardness profile. The measured hardness and grain-size data were used to develop predictive models. The predicted microhardness profiles and grain sizes were then used to explain the effect of machining parameters, such as the cutting-speed, tool-coating and edge-radius upon surface integrity.

Experiments were carried out[180] in order to classify the tool conditions existing during the high-speed machining of Ti-6Al-4V. During machining, vibrational signals were monitored continuously by accelerometer. The features of the signals were analysed and a set of prominent features was selected by using a dimensionality reduction technique. The selected features were then used as inputs to a classification algorithm which judged the condition of the tool. The feature-selection was carried out by using a decision-tree algorithm. Classification of the tool conditions was carried out by using machine-learning algorithms; including a decision-tree algorithm and artificial neural network methods. The latter produced better results.

A study was made[181] of the milling of titanium alloy by using an artificial neural network back-propagation and Levenberg-Marquardt algorithm to correlate the effect of processing parameters. Experiments were performed using a 5-axis milling machine and carbide-cutting inserts of 12 and 14mm under coolant and dry machining conditions. The network supplied a regression analysis which was suitable for the prediction of feasible ranges of process parameters. There was a large reduction in the rate of tool wear under coolant conditions as opposed to dry machining.

The machining of grade-II titanium alloy using minimum-quantity lubrication was studied[182] by using an artificial intelligence method termed cohort intelligence. The cohort intelligence algorithm as used to optimize the processing parameters which are associated with the turning of the above alloy under the stated conditions. The performance of this algorithm was far better when compared with the particle swarm optimization algorithm. A multi cohort intelligence algorithm led to an 8% minimization of the cutting force, a 42% minimization of the tool wear, a 38% minimization of the tool-chip contact length and a 15% minimization of the surface roughness when compared with particle swarm optimisation.

A study was made[183] of the effect of wire electric-discharge machining parameters upon the surface finish of Ti–6Al–4V by using artificial intelligence techniques[184]. The adaptive network based fuzzy inference system (ANFIS) and multi-parametric optimization were used to find the optimum solution for the machining of the titanium superalloy. The peak-current, taper-angle, pulse-on time, pulse-off time and the dielectric fluid flow-rate were selected as operational constraints. The surface roughness was considered to be an output response. The effect upon machining performance was analyzed by using the ANFIS model, and validated using factorial regression models. The models led to a considerable improvement of the process.

Artificial neural network models have been used[185] to predict the surface roughness of Ti-15-3 alloy during electrical discharge machining. The models used the peak current, the pulse on-time, the pulse off-time and the servo-voltage as input parameters. A multilayer perceptron, with 3 hidden layer feed-forward networks was used. Training of the models was performed by using data from an extensive series of experiments with copper electrodes as the positive pole. Predictions which were based upon the trained models were checked by using another set of experiments, and were found to be in good agreement with the experimental results.

Electric discharge machining can be used on workpieces of high hardness, such as titanium alloys and tool steels, but may offer low productivity. In order to improve efficiency, an intelligent system was developed[186] which adaptively controlled debris-removal rather than using pre-set parameters. A feature-extraction method was proposed for the identification of machining states from streaming images of the machining process in order to choose appropriate moments for debris removal. The extracted features were then feed to an artificial neural network model in order to predict debris-removal. Experimental results showed that the model could attain an accuracy of 96.93% over a testing data-set which contained 750 machining images. The debris-removal prediction model was then integrated into an electric discharge machine tool. When compared with pre-set debris-removal machining, the intelligent system could save 38.60% of machining time for a machining depth of 6.45mm.

Electrical discharge hole-drilling, a variant of electrical discharge machining, is used to create starting-holes for wire electrical discharge applications. An automated intelligent system was developed[187] for drilling Inconel-718 and Ti-6Al-4V (Ti64). The system could be used to design the drilling process with regard choosing the optimum settings for parameters such as the discharge current, the pulse on-time, the pulse off-time and the capacitance rate. The total drilling-time, the minimum required electrode-length and the roughness of the resultant surface were the system outputs. The input–output interactions were determined using an adaptive neuro-fuzzy inference technique that permitted design of drilling operations in an efficient and reliable manner.

A new physics-based optimization approach used artificial intelligence to generate digital processing twins[188]. It was applied to the finish-machining of components made from gamma titanium aluminide because such components had suffered from persistent defects such as surface and sub-surface cracks. The fundamental issue was how the cracks occurred during cutting. *In situ* process characterization was combined with machine-learning algorithms to create a model which could reduce the environmental and energy impacts while markedly increasing yield. An improvement in overall energy-efficiency of more than 84%, of 93% in process queuing time, of 2% in scrap cost and of 93% in queuing cost was possible for γ-TiAl machining.

Titanium alloys possess a low thermal conductivity and exhibit great chemical reactivity at high temperatures and are therefore difficult to cut by laser. The process can be optimized by using artificial intelligence methods. A fuzzy logic system was developed[189] in order to predict the laser-cutting behaviour of Ti-6Al-4V sheet. A hybrid neural network and fuzzy logic method was able to predict the kerf-width and kerf-deviation in adequate agreement with experimental data.

Pulsed millisecond Nd:YAG lasers have been used for the micro-drilling of titanium alloys and stainless steels, under identical machining conditions, by varying parameters such as the current, pulse-width, pulse-frequency and gas-pressure. Artificial intelligence techniques such as adaptive neuro-fuzzy inference (ANFIS) and multi-gene genetic programming (MGGP) were used[190] to predict the circularity at entry- and exit-points, the heat-affected zone size, the spatter-area and the taper. Here, 70% of the experimental data made up the training-set while the remaining 30% were used as the testing-set. The root mean square error for the testing data-set ranged from 8.17 to 24.17% and from 4.04 to 18.34% for the ANFIS model and the MGGP model, respectively, when drilling titanium. For the record, the root mean square error for the testing-set ranged from 13.08 to 20.45% and from 6.35 to 10.74% for the ANFIS and MGGP model, respectively, for stainless steel.

The hole-circularity, taper-angle and spatter-area when drilling Ti-6Al-4V using a fibre laser are the basic properties that govern the quality of the product, and these properties are directly related to parameters such as the laser power, the cutting-speed and the gas-pressure. Some 27 experiments were performed[191] using differing parameter-combinations were used to furnish enough data to judge their effect upon product quality. The spatter-fields which were formed in uncontrolled diffused and complex shapes during cutting were calculated by using image-processing. The resultant data were modelled by means of a so-called extreme learning machine based upon artificial intelligence and artificial neural networks. This predicted the hole-diameter, the taper-angle and the spatter-area. A comparison in terms of the modelling performance during the training and testing phases, showed that the extreme learning machine method performed faster than the artificial neural network method and offered a better performance, with smaller error margins.

The prediction of the mechanical properties of welded joints, based upon the use of adaptive fuzzy neural networks was considered[192]. The tensile strength, bend strength and elongation of TC4 joints, welded using the TIG method, were studied. By using back-propagation and hybrid algorithms, the mechanical properties of the welded joints were simulated using various fuzzy sub-sets and training epochs. Use of the hybrid algorithm reduced the average error of training and prediction to less than 7%.

The effect of notch-geometry and temperature upon the tensile properties of titanium alloys was determined[193] by using a Bayesian artificial intelligence model. The model checked whether calculations were reasonable within the context of established solid-mechanics and metallurgical theories. It was possible to gauge the isolated effects of variables such as the elastic stress concentration factor, which cannot be exactly varied independently in practice. This emphasized the ability of such techniques to examine phenomena in those cases where information cannot be accessed experimentally.

A general probabilistic framework was developed[194] which used experimental data as the input, interpolated missing data via plasticity simulation and furnished correlations by means of machine-learning and Bayesian networks. Experimental results on the cycle-by-cycle progress of a short crack growing in a beta-metastable titanium alloy, VST-55531, were gathered by using phase and diffraction contrast tomography. These results were used as inputs for plasticity simulations of the micromechanical fields around the crack. The results demonstrated the predictive capability of the technique in the high-cycle fatigue regime.

Using experimental data on the microstructures, and elemental and fractional compositions, of titanium-alloy powders 4 classes of conformity were identified: properties, optimum properties, possible defects, defective material[195]. These were desirable features of raw materials for additive manufacture. The basic characteristics of a material which earned it a place in a given class were established. Training- and test-samples, based upon 20 features which characterized each of the 4 classes of alloy powder, were used for the learning procedure. The method which was developed for the identification of the class of material was based upon the second-order Kolmogorov-Gabor polynomial and the random forest algorithm. An experimental comparison was made of this method, as compared with random forest, logistic regression and support vector machine methods. The present supervised-learning method permitted the construction of models for processing a large number of each input vector characteristics. Here, the random forest algorithm provided satisfactory generalization while retaining the advantage of an additional increase in accuracy due to the Kolmogorov-Gabor polynomial. The present method permitted an increase in the modelling accuracy of 34.38, 33.34 and 3.13% as compared with those of the support vector machine, logistic regression and random forest, respectively.

The relationship between processing parameters and lack-of-fusion porosity was studied[196] for the laser powder-bed fusion process for Ti-6Al-4V alloy (Ti64). A physics-based thermo-fluid model was used to predict the porosity. An active learning framework was applied to the optimum design of experiments. A customized neural network symbolic regression tool identified a mechanistic relationship between the processing conditions and the porosity. Results indicate that combining the physics-based thermo-fluid model for PBF porosity prediction with active learning and symbolic regression can find an appropriate mechanistic relationship of LOF porosity that is predictive for a wide range of processing conditions.

The effects of laser powder-bed fusion parameters upon the static tensile properties of Ti-6Al-4V were determined[197] by means of artificial intelligence techniques. The parameters included the laser-power, the scanning-speed, the hatch-spacing, the layer-thickness and the sample direction. These were used as the teaching inputs while the yield strength, the ultimate tensile strength and the elongation were the outputs of neural networks. The neural networks included shallow-, deep- and stacked auto-encoder assigned, and were evaluated in terms of their efficiency. A comparison of the accuracies of the outputs of these networks showed that pre-trained stacked auto-encoder assigned deep neural network offered the best performance. The results indicated that increasing the depth of the neural network, in terms of the number of layers, played an important role in improving the accuracy of the predicted outputs. Parametric analysis revealed that laser powder-bed fusion parameters similarly affected the yield and ultimate tensile strengths whereas the elongation exhibited a different trend as a function of the input parameters. It was found that the use of a high laser-power, scanning-speed, hatch-spacing or layer-thickness could have a detrimental effect upon the tensile properties. There existed an optimum range for each of those parameters. The scanning-speed, laser-power, hatch-spacing, layer-thickness and sample-direction played the most significant roles in determining the yield and ultimate tensile strengths … in that order (table 24). In the case of elongation, the laser-power was the most important parameter, while the scanning-speed, the hatch-spacing, the layer-thickness and the sample-direction exerted the least influence, in that order.

Materials Research Forum LLC

https://doi.org/10.21741/9781644903148

Table 24. Effect of laser powder-bed fusion parameters upon the properties of Ti-6Al-4V

Parameter	Property	Relative Effect (%)
scanning speed	UTS	28.0
scanning speed	yield strength	27.5
laser power	UTS	26.0
hatch spacing	UTS	25.0
scanning speed	elongation	24.0
laser power	yield strength	23.0
laser power	elongation	23.0
hatch spacing	yield strength	22.0
hatch spacing	elongation	21.0
layer thickness	elongation	21.0
layer thickness	yield strength	15.0
sample direction	yield strength	13.0
layer thickness	UTS	12.5
sample direction	elongation	12.5
sample direction	UTS	7.5

A hybrid probabilistic neural network and support vector machine high-precision approach was used[198] for the evaluation of alloy properties for the additive manufacturing of biomedical implants[199]. A new approach was used to extend the dimensionality of the input-data space by the outputs of the summation layer of the modified probabilistic neural network topology. Experimental modelling of the approach was carried out by using a data-set of the properties of Ti-6Al-4V and Ti-Al-V-Zr. There was a marked increase in the accuracy of the hybrid scheme, as compared with that of other machine-learning methods.

The low-power fibre laser cutting of Ti-6Al-4V was studied[200] by using a particle-swarm optimization technique. The sawing-angle, power, duty-cycle, pulse-frequency and scanning speed were the input variables for the laser-beam machining process, while the kerf taper and heat-affected zone were output variables.

Figure 28. Construction of L2₁ full-Heusler X₂YZ alloy
Grey spheres: X, checkered spheres: Y, candy-stripe spheres: Z

A cyber-physical quality system was developed[201] which could predict and validate quality-monitoring systems with 95% accuracy in real-time by using machine-learning techniques. Physical data such as the speed, feed, depth-of-cut, coolant temperature, vibrations, tangential cutting forces, and tool life data for 400 parts were collected from sensors on computerized numerical control machines. Machine-learning techniques were used to predict the quality of parts when the inputs affecting them were predominately dominated by vibration and temperature. Extreme gradient boosting machine-learning could predict the quality of a part with 96.2% accuracy. The tool wear was approximated by using Taylor's equation, which could be enhanced by using image processing.

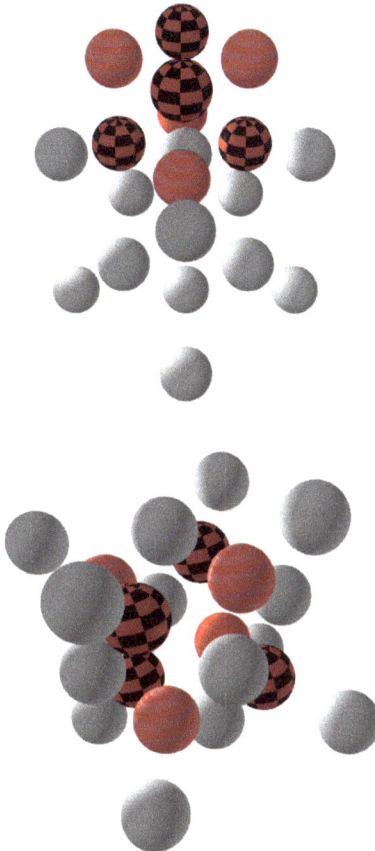

Figure 29. Construction of $C1_b$ half-Heusler XYZ alloy
Grey spheres: X, checkered spheres: Y, candy-stripe spheres: Z

A hybrid artificial intelligence method was used to optimize the atmospheric plasma-spraying parameters for CoMoCrSi coating deposition on Ti-6Al-4V substrates[202]. The Taguchi design method was used to obtain initial solutions for the optimum set of parameters. The true optimum values of the spray-distance, the chamber-pressure, the current, the argon flow-rate and the hydrogen flow-rate were then obtained by using artificial neural networks and genetic algorithms. The coatings which were deposited by using the parameters that were determined by using the Taguchi method alone had a porosity of 8.5% but, following optimization using the algorithms, the porosity was only 5.6%. The as-sprayed coatings contained Cr_3Si as a result of the high-temperature treatment, and the structure of the titanium phase in the coating changed from α-Ti to β-Ti during thermal treatment at up to 1200C. The hardness of the annealed coatings increased with increasing annealing time and annealing temperature. The greater hardness for higher-temperature conditions was attributed to the precipitation of Cr_3Si phase. In general, the coatings following optimum treatments exhibited low porosity, high hardness and good thermal stability at elevated temperatures.

Miscellaneous Alloys

A machine-learning was trained by using previous observations and parameters from physiochemical theories, to guide the high-throughput experimental search for new metallic glasses in the Co-V-Zr ternary system[203]. The experimental observations agreed well with the predictions of the model, but with quantitative discrepancies among the exact compositions which were predicted and which were used to re-train the machine-learning model. The re-trained model offered a much-improved accuracy for the Co-V-Zr, and other, systems. The improved model guided the discovery of metallic glasses in two new ternary systems.

A machine-learning model was trained[204] to find Heusler compounds. These materials were once notable merely for the fact that they were ferromagnetic, although made up of non-ferromagnetic elements. They are now known to exhibit a bewildering range of properties, from semiconducting to superconducting. They exhibit 3 subtly-different crystal structures: Heusler, inverse-Heusler, CsCl-type. They are often difficult to distinguish by using routine diffraction techniques. Another very useful feature is that their properties can often be tailored (to produce a topological insulator, for example) on the basis of simple quantum-mechanical theory (rigid band approach) or predicted by a mere valence-electron count. Some of the alloys are half-metallic ferromagnets; other half-metallic Heusler alloys are compensated ferrimagnets. There has obviously been some considerable departure from the original Heusler formulation, with the alloys nowadays containing ferromagnetic, and even non-metallic, elements. They are now also classed mainly as being either Heusler alloys, having the original 2:1:1 alloying pattern (figure 28), or as Half-Heusler alloys having a 1:1:1 pattern (figure 29). They may also include more than three components, as in the so-called quaternary alloys (figures 30 and 31).

The huge interest in the properties of these alloys, and their ever-increasing ranges of application means that their number is now probably approaching 2000. This is not surprising given that, for the metallurgist, it is very much a pick 'n' mix situation (figure 32). Not only is there this wide choice of constituent elements, there is also little need to be concerned with the usual constrictions of alloy-design. The Hume-Rothery rules still have to be respected, but the familiar need to match other properties is absent because the final characteristics of the alloy depend far more upon a simple valence-electron count than upon the nature and physical properties of the component elements. As compared with other approaches, a so-called Heusler discovery-engine performed could make rapid and reliable predictions of the occurrence of Heusler compounds for an arbitrary combination of elements, given no structural input and over 400000 candidate alloys. The model delivered a true positive rate of 0.94 and a false positive rate of 0.01. It could also detect questionable entries in standard crystallographic databases. It was used to screen candidates which had the form, AB_2C and predict the existence of 12 novel gallides of the form, MRu_2Ga and RuM_2Ga, where M ranged from titanium to cobalt, as being Heusler compounds; predictions which were experimentally confirmed. The material, $TiRu_2Ga$, exhibited superstructure peaks which confirmed that it was an ordered Heusler compound rather than a disordered CsCl-type compound.

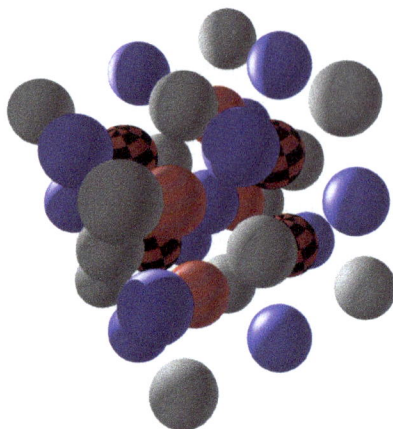

Figure 30. Construction of quaternary Heusler XX'YZ alloy
Light grey spheres: X, dark-grey spheres: X', checkered spheres: Y, candy-stripe
spheres: Z

A support vector regression method, combined with particle swarm optimization and leave-one-out cross-validation was used[205] to construct mathematical methods for the prediction of the magnetic remanence of Nd-Fe-B magnets. The leave-one-out cross-validation of the support vector regression model showed that the mean absolute error did not exceed 0.0036, the mean absolute percentage error was 0.53% and the correlation coefficient could be as high as 0.839.

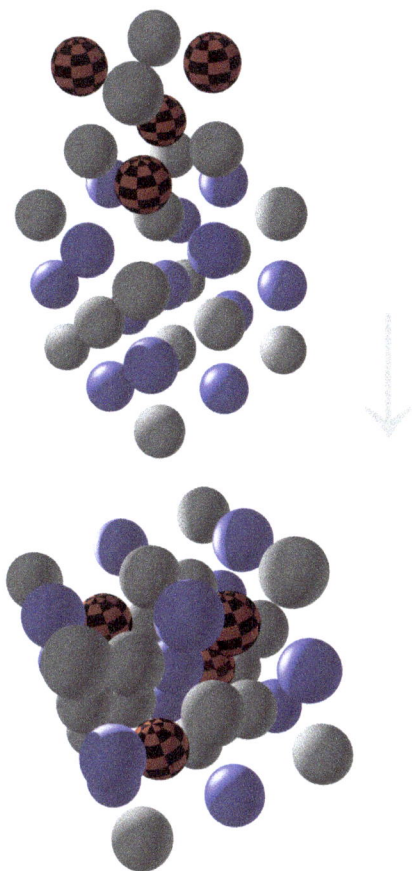

Figure 31. Construction of inverse Heusler X_2YZ alloy
Light grey spheres: X, dark-grey spheres: Y, checkered spheres: Z

The properties of lead-free solders were predicted[206] by using back-propagation neural networks. Various parameters of the neural network were used for training, and the results were analyzed so as to obtain the optimum algorithm and parameter. The effect of adding indium, bismuth, antimony, rare earths and copper to the Sn-Ag alloy upon the tensile strength, shear strength and solidification temperature was considered. The input variables were the latter elements and the outputs were the tensile strength, the shear strength and the solidification temperature. Fifteen groups of data were used to train the model and 3 groups of data were used for prediction. The predicted data were in good agreement with experimental results.

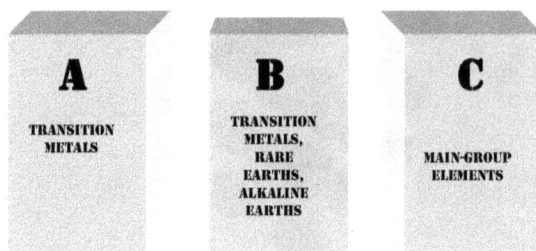

Figure 32a. 'Pick and mix' Full-Heusler alloy-design
Given the basic A_2BC scheme, simply take elements from the relevant box

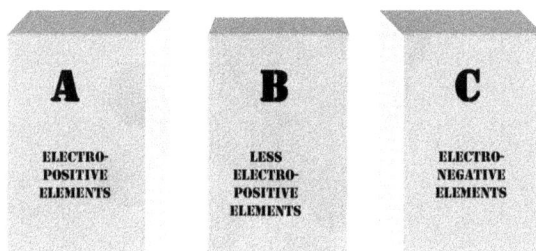

Figure 32b. 'Pick and mix' Half-Heusler alloy-design
Given the basic ABC scheme, simply take elements from the relevant box

The electrical resistivity of electrodeposited $Zn_{1-x}Fe_x$ alloys was investigated[207] as a function of temperature between 10 and 330K for iron concentrations ranging from x = 4 to x = 39 by using genetic programming. There was no established method for predicting the electrical resistivity of electrodeposited alloys as a function of film composition, electrodeposition bath composition and corrosion potential. The object was to develop reliable results, based upon experimental data, and to develop the use of genetic programming for generating predictions of the electrical resistivity. The training and testing sets among a total of 260 samples covered various temperatures and concentrations of components. The training and testing sets consisted of randomly selected 208 and 52, respectively, for the electrical resistivity. The agreement with experimental data was satisfactory.

Given that it is very important to ensure that billets Zircaloy-2 have undergone the proper β-quenching treatment, and have the required β-martensite structure, destructive and semi-destructive techniques such as metallography and hardness measurements have traditionally been used to ensure proper quenching. Ultrasonic velocity and attenuation measurement techniques were later used, with various degrees of success. In this work, logistic regression and support vector machine techniques have now been used[208] to ensure the proper β-quenching treatment of Zircaloy-2 microstructures by using time domain ultrasonic measurements. The results showed that the logistic regression and support vector techniques could both be effective in comparing the microstructures of Zircaloy-2 specimens. The support vector method offered a better confidence level than did logistic regression in ensuring the accurate classification of the microstructures of β-quenched Zircaloy-2.

About the Author

Dr. Fisher has wide knowledge and experience of the fields of engineering, metallurgy and solid-state physics, beginning with work at Rolls-Royce Aero Engines on turbine-blade research, related to the Concord supersonic passenger-aircraft project, which led to a BSc degree (1971) from the University of Wales. This was followed by theoretical and experimental work on the directional solidification of eutectic alloys having the ultimate aim of developing composite turbine blades. This work led to a doctoral degree (1978) from the Swiss Federal Institute of Technology (Lausanne). He then acted for many years as an editor of various academic journals, in particular *Defect and Diffusion Forum*. In recent years he has specialized in writing monographs which introduce readers to the most rapidly developing ideas in the fields of engineering, metallurgy and solid-state physics. He is co-author of the widely-cited student textbook, *Fundamentals of Solidification*. Google Scholar credits him with 8687 citations and a lifetime h-index of 14.

Artificial Intelligence and Alloy Design
Materials Research Foundations **166** (2024)

Materials Research Forum LLC
https://doi.org/10.21741/9781644903148

Keyword Index

References

[1] Iwata S., Ishino S., Mishima Y., IJCAI International Joint Conference on Artificial Intelligence, 1975, 782-788.

[2] Yeh J.W., Chen S.K., Lin S.J., Gan J.Y., Chin T.S., Shun T.T., Tsau C.H., Chang S.Y., Advanced Engineering Materials, 6[5] 2004, 299-303. https://doi.org/10.1002/adem.200300567

[3] Tang B., Kang D., Qing P., Zhang W., Chen N., Journal of Materials Science and Technology, 10[1] 1994, 23-26.

[4] Rychener M.D., Farinacci M.L., Hulthage I., Fox M.S., Proceedings of the National Conference on Artificial Intelligence, 2, 1986, 878-882. https://doi.org/10.1016/0167-8493(86)90006-9

[5] Farinacci M.L., Fox M.S., Hulthage I., Rychener M.D., Robotics, 2[4] 1986, 329-337. https://doi.org/10.1016/0167-8493(86)90006-9

[6] Chen Y.P., Hu D.A., Ma L., Li T.B., Transactions of the China Welding Institution, 21[4] 2000, 20-23.

[7] Jha R., Dulikravich G.S., Chakraborti N., Fan M., Schwartz J., Koch C.C., Colaco M.J., Poloni C., Egorov I.N., Materials and Manufacturing Processes, 32[10] 2017, 1067-1074. https://doi.org/10.1080/10426914.2017.1279319

[8] Gao J., Zhong J., Liu G., Zhang S., Zhang J., Liu Z., Song B., Zhang L., Science and Technology of Advanced Materials, 24[1] 2023, 2196242. https://doi.org/10.1080/14686996.2023.2196242

[9] Jeong S.J., Hwang I.K., Cho I.S., Kim H.S., Journal of Korean Institute of Metals and Materials, 57[3] 2019, 184-192. https://doi.org/10.3365/KJMM.2019.57.3.184

[10] Sunde J.K., Marioara C.D., van Helvoort A.T.J., Holmestad R., Materials Characterization, 142, 2018, 458-469. https://doi.org/10.1016/j.matchar.2018.05.031

[11] Abbod M.F., Talamantes-Silva J., Linkens D.A., Howard I., (2004) Engineering Applications of Artificial Intelligence, 17[5] 2004, 447-456. https://doi.org/10.1016/j.engappai.2004.04.001

[12] Park S., Kayani S.H., Euh K., Seo E., Kim H., Park S., Yadav B.N., Park S.J., Sung H., Jung I.D., Journal of Alloys and Compounds, 903, 2022, 163828. https://doi.org/10.1016/j.jallcom.2022.163828

[13] Gurugubelli S., Chintada V.B., Chekuri R.B.R., International Journal of Vehicle Structures and Systems, 15[2] 2023, 183-186. https://doi.org/10.4273/ijvss.15.2.07

[14] Manish O., Soumen M., Vinay S., Materials Today: Proceedings, 45, 2021, 5069-5073. https://doi.org/10.1016/j.matpr.2021.01.578

[15] Nalci A.S., Cetinkaya M., Karalar A.B., Proceedings of 10th International Conference on Recent Advances in Air and Space Technologies, 2023, in press.

[16] Tiwari T., Jalalian S., Mendis C., Eskin D., Journal of Metals, 2023, in press.

[17] Cao H.P., Zhao X.H., Zhao H., Transactions of the China Welding Institution, 26[2] 2005, 21-24.

[18] Satpathy M.P., Mishra S.B., Sahoo S.K., Journal of Manufacturing Processes, 33, 2018, 96-110. https://doi.org/10.1016/j.jmapro.2018.04.020

[19] Altinkok N., Journal of Composite Materials, 40[9] 2006, 779-796. https://doi.org/10.1177/0021998305055547

[20] Ahamed H., Senthilkumar V., Multidiscipline Modeling in Materials and Structures, 8[2] 2012, 136-158. https://doi.org/10.1108/15736101211251185

[21] Khoshaim A.B., Moustafa E.B., Bafakeeh O.T., Elsheikh A.H., Coatings, 11[12] 2021, 1476. https://doi.org/10.3390/coatings11121476

[22] Rajput R., Raut A., Gangi Setti S., Materials Today: Proceedings, 59, 2022, 1735-1742. https://doi.org/10.1016/j.matpr.2022.04.316

[23] Wang H., Han E., Ke W., Journal of the Chinese Society of Corrosion and Protection, 26[5] 2006, 272-274+281.

[24] Finke A., Escobar J., Munoz J., Petit M., Surface and Coatings Technology, 421, 2021, 127370. https://doi.org/10.1016/j.surfcoat.2021.127370

[25] Nadeau F., Thériault B., Gagné M.O., Proceedings of the Institution of Mechanical Engineers, Part L: Journal of Materials: Design and Applications, 234[5] 2020, 752-765. https://doi.org/10.1177/1464420720917415

[26] Fratini L., Buffa G., Proceedings of the Institution of Mechanical Engineers B, 221[5] 2007, 857-864. https://doi.org/10.1243/09544054JEM674

[27] Maleki E., IOP Conference Series: Materials Science and Engineering, 103[1] 2015, 012034. https://doi.org/10.1088/1757-899X/103/1/012034

[28] Sönmez F., Başak H., Baday S., IDAP 2017 - International Artificial Intelligence and Data Processing Symposium, 2017, 8090325.

[29] Gupta S.K., Pandey K.N., Kumar R., Proceedings of the Institution of Mechanical Engineers L: Journal of Materials: Design and Applications, 232[4] 2018, 333-342. https://doi.org/10.1177/1464420715627293

[30] Shilton A., Rana S., Gupta S., Venkatesh S., 34th Conference on Uncertainty in Artificial Intelligence, 1, 2018, 145-155.

[31] Hartl R., Praehofer B., Zaeh M.F., Proceedings of the Institution of Mechanical Engineers, Part L: Journal of Materials: Design and Applications, 234[5] 2020, 732-751. https://doi.org/10.1177/1464420719899685

[32] Quarto M., Bocchi S., D'Urso G., Giardini C., International Journal of Mechatronics and Manufacturing Systems, 15[2-3], 2022, 149-166. https://doi.org/10.1504/IJMMS.2022.124919

[33] Hartl R., Hansjakob J., Zaeh M.F., International Journal of Advanced Manufacturing Technology, 110[11-12] 2020, 3145-3167. https://doi.org/10.1007/s00170-020-05696-x

[34] Dorbane A., Harrou F., Sun Y., International Conference on Decision Aid Sciences and Applications, DASA 2022, 1553-1557.

[35] Mishra A., Indian Journal of Engineering, 18[49] 2021, 122-133.

[36] Mishra A., Dasgupta A., Forecasting, 4[4] 2022, 787-797. https://doi.org/10.3390/forecast4040043

[37] Köhler T., Schiele M., Glaser M., Schricker K., Bergmann J.P., Augsburg K., Proceedings of the Institution of Mechanical Engineers, Part L: Journal of Materials: Design and Applications, 234[5] 2020, 766-785. https://doi.org/10.1177/1464420720912773

[38] Hu W., Ma Z., Ji S., Qi S., Chen M., Jiang W., Journal of Materials Science and Technology, 53, 2020, 41-52. https://doi.org/10.1016/j.jmst.2020.01.069

[39] Song Q., Ren Z., Ji S., Niu S., Qi W., Chen M., Advanced Engineering Materials, 21[12] 2019, 1900973. https://doi.org/10.1002/adem.201900973

[40] Hu W., Chang X., Ji S., Li F., Song Q., Niu S., Transactions of the China Welding Institution, 41[6] 2020, 54-59, 84.

[41] Mishra A., Dasgupta A., Frattura ed Integrita Strutturale, 16[62] 2022, 448-459. https://doi.org/10.3221/IGF-ESIS.62.31

[42] Arunchai T., Sonthipermpoon K., Apichayakul P., Tamee K., (2015) Rivista Italiana della Saldatura, 67[4] 2015, 473-481.

[43] Hu J., Bi J., Liu H., Li Y., Ao S., Luo Z., Materials, 15[20] 2022, 7323. https://doi.org/10.3390/ma15207323

[44] Dobrzański L.A., Maniara R., Sokolowski J., Kasprzak W., International Journal of Computational Materials Science and Surface Engineering, 1[2] 2007, 214-255. https://doi.org/10.1504/IJCMSSE.2007.014874

[45] Shahani A.R., Setayeshi S., Nodamaie S.A., Asadi M.A., Rezaie S., Journal of Materials Processing Technology, 209[4] 2009, 1920-1935. https://doi.org/10.1016/j.jmatprotec.2008.04.055

[46] Fang S.F., Wang M.P., Song M., Materials and Design, 30[7] 2009, 2460-2467. https://doi.org/10.1016/j.matdes.2008.10.008

[47] Shabani M.O., Mazahery A., Composites B, 45[1] 2013, 185-191. https://doi.org/10.1016/j.compositesb.2012.07.045

[48] Dey S., Sultana N., Kaiser M.S., Dey P., Datta S., Materials and Design, 92, 2016, 522-534. https://doi.org/10.1016/j.matdes.2015.12.076

[49] Saha S., De S., Datta S., IOP Conference Series: Materials Science and Engineering, 912[5] 2020, 052005. https://doi.org/10.1088/1757-899X/912/5/052005

[50] Zhang L., Xu Z., Wei S., Ren X., Wang M., Rare Metal Materials and Engineering, 45[3] 2016, 548-554. https://doi.org/10.1016/S1875-5372(16)30070-4

[51] Sharma N., Khanna R., Singh G., Kumar V., Particulate Science and Technology, 35[6] 2017, 731-741. https://doi.org/10.1080/02726351.2016.1196276

[52] Hayajneh M., Hassan A.M., Alrashdan A., Mayyas A.T., Journal of Alloys and Compounds, 470[1-2] 2009, 584-588. https://doi.org/10.1016/j.jallcom.2008.03.035

[53] Gabsi A.E.H., International Journal on Interactive Design and Manufacturing, 2023, in press.

[54] Marani Barzani M., Zalnezhad E., Sarhan A.A.D., Farahany S., Ramesh S., Measurement: Journal of the International Measurement Confederation, 61, 2015, 150-161. https://doi.org/10.1016/j.measurement.2014.10.003

[55] Eser A., Aşkar Ayyildiz E., Ayyildiz M., Kara F., Advances in Materials Science and Engineering, 2021, 5576600. https://doi.org/10.1155/2021/5576600

[56] Norkey G., Dubey A.K., Agrawal S., Journal of Intelligent and Fuzzy Systems, 27[3] 2014, 1545-1555. https://doi.org/10.3233/IFS-141121

[57] Hu Y., Xie J., Liu Z., Ding Q., Zhu W., Zhang J., Zhang W., Computational Materials Science, 142, 2018, 244-254. https://doi.org/10.1016/j.commatsci.2017.09.059

[58] Nalajam P.K., Micron, 151, 2021, 103161. https://doi.org/10.1016/j.micron.2021.103161

[59] Ren W., Mazumder J., Scientific Reports, 10[1] 2020, 19493. https://doi.org/10.1038/s41598-020-75131-4

[60] Muhammad W., Brahme A.P., Ibragimova O., Kang J., Inal K., International Journal of Plasticity, 136, 2021, 102867. https://doi.org/10.1016/j.ijplas.2020.102867

[61] Chheda A.M., Nazro L., Sen F.G., Hegadekatte V., IOP Conference Series: Materials Science and Engineering, 651[1] 2019, 012107. https://doi.org/10.1088/1757-899X/651/1/012107

[62] Affronti E., Jaremenko C., Merklein M., Maier A., Materials, 11[9] 2018, 1495. https://doi.org/10.3390/ma11091495

[63] Jaremenko C., Affronti E., Merklein M., Maier A., Materials, 13[11] 2020, 2427. https://doi.org/10.3390/ma13112427

[64] Dong F., Xu W., Wu Z., Hou J., Steel and Composite Structures, 47[5] 2023, 601-613.

[65] Bobbili R., Ramakrishna B., Madhu V., Mechanics Based Design of Structures and Machines, 51[1] 2023, 327-338. https://doi.org/10.1080/15397734.2020.1843487

[66] Varol Ö.H., İnce M., Bıçaklı E.E., Arabian Journal for Science and Engineering, 48[3] 2023, 2841-2850. https://doi.org/10.1007/s13369-022-07009-8

[67] Yasnii O.P., Pastukh O.A., Pyndus Y.I., Lutsyk N.S., Didych I.S., Materials Science, 54[3] 2018, 333-338. https://doi.org/10.1007/s11003-018-0189-9

[68] Zafar M.H., Younis H.B., Mansoor M., Moosavi S.K.R., Khan N.M., Akhtar N., Materials, 15[18] 2022, 6198. https://doi.org/10.3390/ma15186198

[69] Abdullatef M.S., Alzubaidi F.N., Al-Tamimi A., Mahmood Y.A., Fluid Dynamics and Materials Processing, 19[8] 2023, 2083-2083. https://doi.org/10.32604/fdmp.2023.027266

[70] Gupta A.K., Kumar S., Chandna P., Bhushan G., Silicon, 13[8] 2021, 2429-2443. https://doi.org/10.1007/s12633-020-00594-z

[71] Oyekunle F., Abou-El-Hossein K., Lecture Notes in Mechanical Engineering, 2021, 191-200. https://doi.org/10.1007/978-981-15-9893-7_13

[72] Samanta B., International Journal of Computer Integrated Manufacturing, 22[3] 2009, 257-266. https://doi.org/10.1080/09511920802287138

[73] Kanna S.K.R., Sundar G.N., Ganesan R., Mallireddy N., Shingadia H., Anandaram H., De Poures M.V., Paramasivam P., Journal of Nanomaterials, 2022, 2892738. https://doi.org/10.1155/2022/2892738

[74] Kim H., Lee S.H., IEEE Access, 10, 2022, 67826-67838. https://doi.org/10.1109/ACCESS.2022.3186336

[75] Kordijazi A., Behera S., Patel D., Rohatgi P., Nosonovsky M., Langmuir, 37[12] 2021, 3766-3777. https://doi.org/10.1021/acs.langmuir.1c00358

[76] Gusel L., Brezocnik M., Rudolf R., Anzel I., Lazarević Z., Romčević N., Optoelectronics and Advanced Materials, Rapid Communications, 4[3] 2010, 395-400.

[77] Fang S.F., Wang M.P., Materials Science Forum, 789, 2014, 574-579. https://doi.org/10.4028/www.scientific.net/MSF.789.574

[78] Ozdemir R., Karahan I.H., Journal of Optoelectronics and Advanced Materials, 17[1-2] 2015, 14-26.

[79] El-Sawy A., P. P. Abdul Majeed A., Musa R.M., Mohd Razman M.A., Hassan M.H.A., Jaafar A.A., Lecture Notes in Mechanical Engineering, 2020, 403-407. https://doi.org/10.1007/978-981-13-8323-6_34

[80] Deng Z., Yin H., Jiang X., Zhang C., Zhang K., Zhang T., Xu B., Zheng Q., Qu X., Computational Materials Science, 155, 2018, 48-54. https://doi.org/10.1016/j.commatsci.2018.07.049

[81] Sivam S.P.S.S., Harshavardhana N., Rajendran R., Proceedings of the Institution of Mechanical Engineers, Part C: Journal of Mechanical Engineering Science, 2023, in press.

[82] Islam N., Huang W., Zhuang H.L., Computational Materials Science, 150, 2018, 230-235. https://doi.org/10.1016/j.commatsci.2018.04.003

[83] Khatamsaz D., Vela B., Singh P., Johnson D.D., Allaire D., Arróyave R., Acta Materialia, 236, 2022, 118133. https://doi.org/10.1016/j.actamat.2022.118133

[84] Lee K., Balachandran P.V., Materialia, 26, 2022, 101628. https://doi.org/10.1016/j.mtla.2022.101628

[85] Agarwal A., Prasada Rao A.K., Journal of Metals, 71[10] 2019, 3424-3432. https://doi.org/10.1007/s11837-019-03712-4

[86] Lei K., Joress H., Persson N., Hattrick-Simpers J.R., DeCost B., Journal of Chemical Physics, 155[5] 2021, 054105. https://doi.org/10.1063/5.0050885

[87] Li X., Li B., Yang Z., Chen Z., Gao W., Jiang Q., Journal of Materials Chemistry A, 10[2] 2022, 872-880. https://doi.org/10.1039/D1TA09184K

[88] Li S., Li S., Liu D., Zou R., Yang Z., Computational Materials Science, 205, 2022, 111185. https://doi.org/10.1016/j.commatsci.2022.111185

[89] Bobbili R., Ramakrishna B., Materials Today Communications, 36, 2023, 106674. https://doi.org/10.1016/j.mtcomm.2023.106674

[90] Jain R., Lee U., Samal S., Park N., Journal of Alloys and Compounds, 956, 2023, 170193. https://doi.org/10.1016/j.jallcom.2023.170193

[91] Zhang Y.F., Ren W., Wang W.L., Li N., Zhang Y.X., Li X.M., Li W.H., Journal of Alloys and Compounds, 945, 2023, 169329. https://doi.org/10.1016/j.jallcom.2023.169329

[92] Reddy N.S., Krishnaiah J., Kumar Y.K., Acharya N.N., Proceedings of the 9th International Conference on Neural Information Processing: Computational Intelligence for the E-Age, 2, 2002, 829-833.

[93] Chong Z.S., Wilcox S., Ward J., International Journal of COMADEM, 9[3] 2006, 7-14.

[94] Chong Z.S., Wilcox S., Ward J., Proceedings of the ASME International Design Engineering Technical Conferences and Computers and Information in Engineering Conference, 2005, 607-613.

[95] Lin Y.C., Zhang J., Zhong J., Computational Materials Science, 43[4] 2008, 752-758. https://doi.org/10.1016/j.commatsci.2008.01.039

[96] Mirzadeh H., Najafizadeh A., Materials Characterization, 59[11] 2008, 1650-1654. https://doi.org/10.1016/j.matchar.2008.03.004

[97] Bobbili R., Materials Letters, 349, 2023, 134774. https://doi.org/10.1016/j.matlet.2023.134774

[98] Chen T.C., Rajiman R., Elveny M., Guerrero J.W.G., Lawal A.I., Dwijendra N.K.A., Surendar A., Danshina S.D., Zhu Y., Arabian Journal for Science and Engineering, 46[12] 2021, 12417-12425. https://doi.org/10.1007/s13369-021-05966-0

[99] Lee M.W., Choi Y.S., Kwon D.H., Cha E.J., Kang H.B., Jeong J.I., Lee S.J., Kim H.J., Archives of Metallurgy and Materials, 67[4] 2022, 1539-1542. https://doi.org/10.24425/amm.2022.141090

[100] Djurabekova F.G., Domingos R., Cerchiara G., Castin N., Vincent E., Malerba L., Nuclear Instruments and Methods in Physics Research B, 255, 2007, 8-12. https://doi.org/10.1016/j.nimb.2006.11.039

[101] Castin N., Malerba L., Bonny G., Pascuet M.I., Hou M., Nuclear Instruments and Methods in Physics Research B, 267[18] 2009, 3002-3008. https://doi.org/10.1016/j.nimb.2009.06.092

[102] Pascuet M.I., Castin N., Becquart C.S., Malerba L., Journal of Nuclear Materials, 412[1] 2011, 106-115. https://doi.org/10.1016/j.jnucmat.2011.02.038

[103] Anirudh M.K., Iyengar M.S., Desik P.H.A., Phaniraj M.P., Oxidation of Metals, 98, 2022, 291-303. https://doi.org/10.1007/s11085-022-10123-5

[104] Ravi S.K., Roy I., Roychowdhury S., Feng B., Ghosh S., Reynolds C., Umretiya R.V., Rebak R.B., Hoffman A.K., Computational Materials Science, 230, 2023, 112440. https://doi.org/10.1016/j.commatsci.2023.112440

[105] Jiménez-Come M.J., Muñoz E., García R., Matres V., Martín M.L., Trujillo F., Turias I., Journal of Applied Logic, 10[4] 2012, 291-297. https://doi.org/10.1016/j.jal.2012.07.005

[106] Urda D., Luque R.M., Jiménez M.J., Turias I., Franco L., Jerez J.M., Lecture Notes in Computer Science (including sub-series Lecture Notes in Artificial Intelligence and Lecture Notes in Bioinformatics), 7902[1] 2013, 88-95. https://doi.org/10.1007/978-3-642-38679-4_7

[107] Lu Q., Xu W., Van Der Zwaag S., Proceedings of the International Conference on Solid-Solid Phase Transformations in Inorganic Materials, 2015, 791-792.

[108] Rachkov V.I., Obraztsov S.M., Konobeev Yu.V., Solovev V.A., Belomyttsev M.Y., Molyarov A.V., Atomic Energy, 116[5] 2014, 311-314. https://doi.org/10.1007/s10512-014-9858-4

[109] Khalaj O., Jamshidi M.B., Saebnoori E., Masek B., Stadler C., Svoboda J., IEEE Access, 9, 2021, 156930-156946. https://doi.org/10.1109/ACCESS.2021.3129454

[110] Iqbal A., Arabian Journal for Science and Engineering, 39[11] 2014, 8253-8263. https://doi.org/10.1007/s13369-014-1402-2

[111] Alajmi M.S., Almeshal A.M., Applied Sciences, 11[9] 2021, 4055. https://doi.org/10.3390/app11094055

[112] Veic M., Bajic D., Jozic S., Mechanical Technology and Structural Materials, 2017[55] 2017, 149-158.

[113] Contini L., Jr., Balancin O., Materials Research, 25, 2022, e20220075. https://doi.org/10.1590/1980-5373-mr-2022-0075

[114] Ohno M., Kimura D., Matsuura K., Journal of the Iron and Steel Institute of Japan, 103[12] 2017, 720-729. https://doi.org/10.2355/tetsutohagane.TETSU-2017-040

[115] Ohno M., Kimura D., Matsuura K., Journal of the Iron and Steel Institute of Japan, 103[12] 2017, 711-719. https://doi.org/10.2355/tetsutohagane.TETSU-2017-028

[116] Chaudhary N., Abu-Odeh A., Karaman I., Arróyave R., Journal of Materials Science, 52[18] 2017, 11048-11076. https://doi.org/10.1007/s10853-017-1252-x

[117] Moreira L.C., Li W.D., Lu X., Fitzpatrick M.E., Computers and Industrial Engineering, 127, 2019, 158-168. https://doi.org/10.1016/j.cie.2018.12.016

[118] Rahnama A., Clark S., Sridhar S., Computational Materials Science, 154, 2018, 169-177. https://doi.org/10.1016/j.commatsci.2018.07.055

[119] Mulewicz B., Korpala G., Kusiak J., Prahl U., Materials Science Forum, 949, 2019, 24-31. https://doi.org/10.4028/www.scientific.net/MSF.949.24

[120] Keprate A., Ratnayake R.M.C., Proceedings of the International Conference on Offshore Mechanics and Arctic Engineering, 2020, V003T03A017.

[121] Mrzygłód B., Gumienny G., Wilk-Kołodziejczyk D., Regulski K., Journal of Materials Engineering and Performance, 28[7] 2019. 3894-3904. https://doi.org/10.1007/s11665-019-03932-4

[122] Brili N., Ficko M., Klančnik S., Sensors, 21[5] 2021, 1-18. https://doi.org/10.3390/s21051917

[123] Alajmi M.S., Almeshal A.M., Materials, 14[14] 2021, 3773. https://doi.org/10.3390/ma14143773

[124] Na J., Kim G., Kang S.H., Kim S.J., Lee S., Acta Materialia, 214, 2021, 116987. https://doi.org/10.1016/j.actamat.2021.116987

[125] Drissi-Daoudi R., Pandiyan V., Logé R., Shevchik S., Masinelli G., Ghasemi-Tabasi H., Parrilli A., Wasmer K., Virtual and Physical Prototyping, 17[2] 2022, 181-204. https://doi.org/10.1080/17452759.2022.2028380

[126] Ahmed W., Hegab H., Mohany A., Kishawy H., Materials, 14[20] 2021, 6106. https://doi.org/10.3390/ma14206106

[127] Liu P., Song Y., Chai M., Han Z., Zhang Y., Materials, 14[24] 2021, 7504. https://doi.org/10.3390/ma14247504

[128] Jeon J., Seo N., Jung J.G., Kim H.S., Son S.B., Lee S.J., Journal of Materials Research and Technology, 21, 2022. 1408-1418. https://doi.org/10.1016/j.jmrt.2022.09.119

[129] Nguyen D.N., Kino H., Miyake T., Dam H.C., MRS Bulletin, 48[1] 2023, 31-44. https://doi.org/10.1557/s43577-022-00372-9

[130] Gélinas S., Blais C., Powder Metallurgy, 66[1] 2023, 29-42. https://doi.org/10.1080/00325899.2022.2055888

[131] Sun W.H., Jiao J.C., Li S.M., Cui J., Cao J.S., Wang M., Iron and Steel, 57[8] 2022, 168-176.

[132] Rawa M.J.H., Razavi Dehkordi M.H., Kholoud M.J., Abu-Hamdeh N.H., Azimy H., Engineering Applications of Artificial Intelligence, 126, 2023, 107025. https://doi.org/10.1016/j.engappai.2023.107025

[133] Gao Z., Fan X., Xia T., Xue W., Gao S., Journal of Physics: Conference Series, 2459[1] 2023, 012139. https://doi.org/10.1088/1742-6596/2459/1/012139

[134] Pattanayak S., Dey S., Chatterjee S., Chowdhury S.G., Datta S., Computational Materials Science, 104, 2015, 60-68. https://doi.org/10.1016/j.commatsci.2015.03.029

[135] Fragassa C., Babic M., Pavlovic A., Proceedings on Engineering Sciences, 1[1] 2019, 418-422. https://doi.org/10.24874/PES01.01.054

[136] Nasser M., Attyaoui S., Tlili B., Montagne A., Briki J., Iost A., Materials Research Express, 8[11] 2021, A19. https://doi.org/10.1088/2053-1591/ac3041

[137] Park S.M., Lee T., Lee J.H., Kang J.S., Kwon M.S., Journal of Materials Research and Technology, 22, 2023, 3294-3301. https://doi.org/10.1016/j.jmrt.2022.12.171

[138] Choi Y.S., Kwon D.H., Lee M.W., Cha E.J., Jeon J., Lee S.J., Kim J., Kim H.J., Archives of Metallurgy and Materials, 67[4] 2022, 1459-1463. https://doi.org/10.24425/amm.2022.141074

[139] Kong B.O., Kim M.S., Kim B.H., Lee J.H., Metals and Materials International, 29[5] 2023, 1334-1345. https://doi.org/10.1007/s12540-022-01312-7

[140] Gui Y., Li Q., Zhu K., Xue Y., Materials Today Communications, 27, 2021, 102282. https://doi.org/10.1016/j.mtcomm.2021.102282

[141] Jafari M., Lotfi M., Ghaseminejad P., Roodi M., Teimouria R., Transactions of the Indian Institute of Metals, 68[5] 2015, 969-979. https://doi.org/10.1007/s12666-015-0535-7

[142] Wiese B., Berger S., Bohlen J., Nienaber M., Höche D., Materials Today Communications, 36, 2023, 106566. https://doi.org/10.1016/j.mtcomm.2023.106566

[143] Sharma R., Chelladurai I., Orme A.D., Miles M.P., Giraud-Carrier C., Fullwood D.T., Journal of Microscopy, 272[1] 2018, 67-78. https://doi.org/10.1111/jmi.12748

[144] Yadav P.K., Khurana M.K., Proceedings of the Institution of Mechanical Engineers, Part L: Journal of Materials: Design and Applications, 236[2] 2022, 345-360. https://doi.org/10.1177/14644207211044805

[145] Jagadeesh G.V., Setti S.G., Transactions of the Indian Institute of Metals, 75[6] 2022, 1603-1618. https://doi.org/10.1007/s12666-022-02536-2

[146] Manroo S.A., Malik V., Transactions of the Indian Institute of Metals, 75[8] 2022, 2051-2059. https://doi.org/10.1007/s12666-022-02581-x

[147] Mishra A., International Journal on Interactive Design and Manufacturing, 2023,in preparation.

[148] Corne R., Nath C., Mansori M.E., Kurfess T., Procedia Manufacturing, 5, 2016, 1-14. https://doi.org/10.1016/j.promfg.2016.08.004

[149] Cica D., Kramar D., Artificial Intelligence: Advances in Research and Applications, 2017, 147-169.

[150] Elsheikh A.H., Muthuramalingam T., Shanmugan S., Mahmoud Ibrahim A.M., Ramesh B., Khoshaim A.B., Moustafa E.B., Bedairi B., Panchal H., Sathyamurthy R., Journal of Materials Research and Technology, 15, 2021, 3622-3634. https://doi.org/10.1016/j.jmrt.2021.09.119

[151] Thejasree P., Narasimhamu K.L., Natarajan M., Raju R., International Journal on Interactive Design and Manufacturing, 2022, in press.

[152] Mrzygłód B., Łukaszek-Sołek A., Olejarczyk-Wożeńska I., Pasierbiewicz K., Archives of Foundry Engineering, 22[3] 2022, 41-52. https://doi.org/10.24425/afe.2022.140235

[153] Jiang X., Yin H.Q., Zhang C., Zhang R.J., Zhang K.Q., Deng Z.H., Liu G.Q., Qu X.H., Computational Materials Science, 143, 2018, 295-300. https://doi.org/10.1016/j.commatsci.2017.09.061

[154] Jun J., Shin D., Dryepondt S., Haynes J.A., Pint B.A., NACE - International Corrosion Conference Series, 2018.

[155] Nandal V., Dieb S., Bulgarevich D.S., Osada T., Koyama T., Minamoto S., Demura M., Scientific Reports, 13[1] 2023, 12660. https://doi.org/10.1038/s41598-023-39589-2

[156] Chandran M., Lee S.C., Shim J.H., Modelling and Simulation in Materials Science and Engineering, 26[2] 2018, 025010. https://doi.org/10.1088/1361-651X/aa9f37

[157] Catal A.A., Bedir E., Yilmaz R., Canadinc D., Journal of Alloys and Compounds, 904, 2022, 164135. https://doi.org/10.1016/j.jallcom.2022.164135

[158] Lu B., Wang C., Chinese Physics B, 27[7] 2018, 077104. https://doi.org/10.1088/1674-1056/27/7/077104

[159] Moon I.Y., Jeong H.W., Lee H.W., Kim S.J., Oh Y.S., Jung J., Oh S., Kang S.H., Metals, 12[2] 2022, 223. https://doi.org/10.3390/met12020223

[160] Ma B., Wang X., Xu G., Xu J., He J., Materials, 15[19] 2022, 6523. https://doi.org/10.3390/ma15196523

[161] Jha R., Dulikravich G.S., Metals, 11[1] 2021, 1-15. https://doi.org/10.3390/met11010015

[162] Li P., Xue K.M., Chinese Journal of Nonferrous Metals, 16[7] 2006, 1202-1206.

[163] Wang L.Y., Li L., Zhang Z.H., Journal of Materials Engineering and Performance, 25[9] 2016, 3912-3923. https://doi.org/10.1007/s11665-016-2230-1

[164] Deb S., Muraleedharan A., Immanuel R.J., Panigrahi S.K., Racineux G., Marya S., Theoretical and Applied Fracture Mechanics, 119, 2022, 103338. https://doi.org/10.1016/j.tafmec.2022.103338

[165] Ramesh S., Karunamoorthy L., Palanikumar K., Materials and Manufacturing Processes, 23[4] 2008, 439-447. https://doi.org/10.1080/10426910801976676

[166] D'Mello G., Pai P.S., Puneet N.P., Applied Soft Computing Journal, 51, 2017, 105-115. https://doi.org/10.1016/j.asoc.2016.12.003

[167] Muthuram N., Frank F.C., Materials Today: Proceedings, 46, 2021, 8097-8102. https://doi.org/10.1016/j.matpr.2021.03.045

[168] Kim T., Kim J.G., Park S., Kim H.S., Kim N., Ha H., Choi S.K., Tucker C., Sung H., Jung I.D., Virtual and Physical Prototyping, 18[1] 2023, e2124921. https://doi.org/10.1080/17452759.2022.2124921

[169] Akhil V., Raghav G., Arunachalam N., Srinivas D.S., Journal of Computing and Information Science in Engineering, 20[2] 2020, 021010. https://doi.org/10.1115/1.4045719

[170] Krishnakumar P., Rameshkumar K., Ramachandran K.I., Intelligent Decision Technologies, 12[2] 2018, 265-282. https://doi.org/10.3233/IDT-180332

[171] Krishnakumar P., Rameshkumar K., Ramachandran K.I., International Journal of Computational Intelligence and Applications, 17[3] 2018, 1850017. https://doi.org/10.1142/S1469026818500177

[172] Escamilla I., Torres L., Perez P., Zambrano P., Lecture Notes in Computer Science (including subseries Lecture Notes in Artificial Intelligence and Lecture Notes in Bioinformatics), 5317, 2008, 1009-1019. https://doi.org/10.1007/978-3-540-88636-5_95

[173] Roushan A., Srinivas Rao U., Patra K., Sahoo P., Journal of Physics: Conference Series, 1950[1] 2021, 012046. https://doi.org/10.1088/1742-6596/1950/1/012046

[174] Thepsonthi T., Ozel T., Transactions of the North American Manufacturing Research Institution of SME, 42, 2014, 158-167.

[175] Al-Zubaidi S., Ghani J.A., Che Haron C.H., Modelling and Simulation in Engineering, 2013, 932094. https://doi.org/10.1155/2013/932094

[176] Al-Zubaidi S., Ghani J.A., Che Haron C.H., Arabian Journal for Science and Engineering, 39[6] 2014, 5095-5111. https://doi.org/10.1007/s13369-014-0975-0

[177] Nasr M.M., Anwar S., Al-Samhan A.M., Ghaleb M., Dabwan A., Materials, 13[24] 2020, 1-22.

[178] Caggiano A., Sensors, 18[3] 2018, 823. https://doi.org/10.3390/s18030823

[179] Arisoy Y.M., Özel T., Materials and Manufacturing Processes, 30[4] 2015, 425-433. https://doi.org/10.1080/10426914.2014.961476

[180] Krishnakumar P., Rameshkumar K., Ramachandran K.I., Procedia Computer Science, 50, 2015, 270-275. https://doi.org/10.1016/j.procs.2015.04.049

[181] Daniyan I.A., Mpofu K., Tlhabadira I., Ramatsetse B.I., International Journal of Mechanical Engineering and Robotics Research, 10[11] 2021, 601-611. https://doi.org/10.18178/ijmerr.10.11.601-611

[182] Shastri A., Nargundkar A., Kulkarni A.J., Benedicenti L., SN Applied Sciences, 3[2] 2021, 226. https://doi.org/10.1007/s42452-021-04197-0

[183] Kumar S., Dhanabalan S., Narayanan C.S., International Journal of Decision Support System Technology, 11[4] 2019, 96-115. https://doi.org/10.4018/IJDSST.2019100105

[184] Kumar S., Dhanabalan S., Narayanan C.S., SN Applied Sciences, 1[4] 2019, 298. https://doi.org/10.1007/s42452-019-0195-z

[185] Khan M.A.R., Rahman M.M., Kadirgama K., Maleque M.A., Bakar R.A., World Academy of Science, Engineering and Technology, 74, 2011, 194-202.

[186] Lee C.H., Lai T.S., IEEE Access, 9, 2021, 75302-75312. https://doi.org/10.1109/ACCESS.2021.3080297

[187] Yilmaz O., Bozdana A.T., Okka M.A., International Journal of Advanced Manufacturing Technology, 74[9-12] 2014, 1323-1336. https://doi.org/10.1007/s00170-014-6059-1

[188] Adeniji D., Schoop J., Proceedings of the ASME 2021 16th International Manufacturing Science and Engineering Conference, MSEC 2021, 1.

[189] Pandey A.K., Dubey A.K., Machining Science and Technology, 17[4] 2013, 545-574. https://doi.org/10.1080/10910344.2013.806182

[190] Chatterjee S., Mahapatra S.S., Bharadwaj V., Upadhyay B.N., Bindra K.S., Engineering with Computers, 37[2] 2021, 1181-1204. https://doi.org/10.1007/s00366-019-00878-y

[191] Ay M., Journal of Manufacturing Processes, 36, 2018, 138-148. https://doi.org/10.1016/j.jmapro.2018.10.003

[192] Dong J., Zhang Y., Tang Z., Transactions of the China Welding Institution, 29[7] 2008, 29-33.

[193] Das A., International Journal of Materials Research, 109[11] 2018, 979-1004.

[194] Rovinelli A., Guilhem Y., Proudhon H., Lebensohn R.A., Ludwig W., Sangid M.D., Modelling and Simulation in Materials Science and Engineering, 25[4] 2017, 045010. https://doi.org/10.1088/1361-651X/aa6c45

[195] Tkachenko R., Duriagina Z., Lemishka I., Izonin I., Trostianchyn A., Eastern-European Journal of Enterprise Technologies, 3[12-93] 2018, 23-31. https://doi.org/10.15587/1729-4061.2018.134319

[196] Mojumder S., Gan Z., Li Y., Amin A.A., Liu W.K., Additive Manufacturing, 68, 2023, 103500. https://doi.org/10.1016/j.addma.2023.103500

[197] Maleki E., Bagherifard S., Guagliano M., International Journal of Mechanics and Materials in Design, 18[1] 2022, 199-222. https://doi.org/10.1007/s10999-021-09570-w

[198] Izonin I., Tkachenko R., Gregus M., Duriagina Z., Shakhovska N., Computers, Materials and Continua, 71[2], 2022, 5933-5947. https://doi.org/10.32604/cmc.2022.022582

[199] Izonin I., Tkachenko R., Duriagina Z., Shakhovska N., Kovtun V., Lotoshynska N., Applied Sciences, 12[10] 2022, 5238. https://doi.org/10.3390/app12105238

[200] Pramanik D., Roy N., Kuar A.S., Sarkar S., Mitra S., Optics and Laser Technology, 147, 2022, 107613. https://doi.org/10.1016/j.optlastec.2021.107613

[201] Chacko M., Atul, Boddapati S.B., Jordan Journal of Mechanical and Industrial Engineering, 16[5] 2022, 821-833.

[202] Lin C.M., Yen S.H., Su C.Y., Measurement: Journal of the International Measurement Confederation, 94, 2016, 157-167. https://doi.org/10.1016/j.measurement.2016.07.077

[203] Ren F., Ward L., Williams T., Laws K.J., Wolverton C., Hattrick-Simpers J., Mehta A., Science Advances, 4[4] 2018, eaaq1566. https://doi.org/10.1126/sciadv.aaq1566

[204] Oliynyk A.O., Antono E., Sparks T.D., Ghadbeigi L., Gaultois M.W., Meredig B., Mar A., Chemistry of Materials, 28[20] 2016, 7324-7331. https://doi.org/10.1021/acs.chemmater.6b02724

[205] Cheng W., Proceedings of IEEE 13th International Conference on Cognitive Informatics and Cognitive Computing, 2014, 431-435.

[206] Sun L., Qi F.J., Hou Z.Z., Qin C., Journal of Shanghai Jiaotong University, 13, 2008, 138-140. https://doi.org/10.1007/s12204-008-0110-z

[207] Karahan I.H., Ozdemir R., Optoelectronics and Advanced Materials, Rapid Communications, 4[6] 2010, 812-815.

[208] Ramkishore S., Madhumitha P., Palanichamy P., Proceedings - International Conference on Soft Computing and Machine Intelligence, 2014, 130-134.

www.ingramcontent.com/pod-product-compliance
Lightning Source LLC
Chambersburg PA
CBHW071710210326
41597CB00017B/2426